THE STORY OF THE PLANT
KINGDOM

THE UNIVERSITY OF CHICAGO PRESS
CHICAGO, ILLINOIS

—

THE BAKER & TAYLOR COMPANY
NEW YORK

THE CAMBRIDGE UNIVERSITY PRESS
LONDON

THE MARUZEN-KABUSHIKI-KAISHA
TOKYO, OSAKA, KYOTO, FUKUOKA, SENDAI

THE COMMERCIAL PRESS, LIMITED
SHANGHAI

THE
STORY OF THE PLANT
KINGDOM

By

MERLE C. COULTER

Professor of Botany, The University of Chicago

THE UNIVERSITY OF CHICAGO PRESS
CHICAGO · ILLINOIS

COPYRIGHT 1935 BY THE UNIVERSITY OF CHICAGO
ALL RIGHTS RESERVED. PUBLISHED OCTOBER 1935
SEVENTH IMPRESSION JANUARY 1942

COMPOSED AND PRINTED BY THE UNIVERSITY OF CHICAGO PRESS
CHICAGO, ILLINOIS, U.S.A.

PREFACE

FOUR years ago the University of Chicago adopted a new college curriculum. It was hoped that this curriculum would succeed more fully than had its predecessors in "rounding out a respectable minimum of general education" during the Freshman and Sophomore years. Today it appears that the original hope has been realized to a gratifying degree.[1] The University of Chicago proposes, therefore, to continue with her new curriculum. Curricula with similar features are being adopted increasingly in other colleges and junior colleges.

At Chicago the core of the "respectable minimum of general education" is provided by four one-year courses, "Introduction to the Humanities," "Introduction to the Social Sciences," "Introduction to the Physical Sciences," and "Introduction to the Biological Sciences." Local parlance refers to these as "comprehensive courses," as "survey courses," as "orientation courses," and as, simply, "introductory courses." The course instructors wince when the first two titles are used. They recognize frankly that their courses fall far short of being comprehensive. They claim, or at least hope, that their courses are not characterized by the extreme superficiality that is so commonly attributed to "survey courses." The "orientation" title properly suggests that these courses have value in enabling the student to make a fairly intelligent choice of his career, but is in danger of implying—improperly—that such is the sole or the main value of the courses. "Introductory course" is apparently the least offensive of the titles, for the instructors feel that their major purpose is to provide that type of introduction into each of these four fields of human thought that is an appropriate part of the general education of students of college caliber.

For each of these courses there is a syllabus which furnishes out-

---

[1] A description of this curriculum, its operation, and some of its results appears in *The Chicago College Plan*, written by Dean C. S. Boucher and published by the University of Chicago Press.

lines and brief descriptions of the subject matter that is to be studied, together with reading references. Certain chapters in certain books are assigned as "indispensable readings" for all students; while other chapters in the same books as well as selections from other books are cited as "optional readings" for the benefit of such students as may care to delve further into certain topics. In the Introductory Biology Course all "indispensable readings" are contained in a group of ten books. The present volume has been prepared to take its place as one of that group of ten.

Of the twenty chapters which follow, twelve have been prepared to constitute "indispensable readings" on the plant kingdom (chaps. i, ii, iv, v, vi, viii, x, xi, xiii, xiv, xvii, xviii). These may be read intelligibly without reference to the others. In these twelve chapters the author has attempted to describe, and occasionally to explain, certain essentials of botanical science in a style which minimizes technical terminology and detail. Simplicity has been sought for two reasons: (1) the larger part of the audience, having no plan to carry any further the study of botany, is interested only in such botany as constitutes an appropriate part of general education; (2) of the ten volumes of "indispensable readings" this one is encountered earliest in the year.

As a major framework of organization the author has employed the phylogenetic sequence. This fits into the organization of the entire course, in which a phylogenetic survey of plant and animal kingdoms precedes a more penetrating, analytical study of the functions and environmental relationships of living organisms.

Though prepared to fit into this particular niche in the college curriculum at Chicago, the book might also prove of some value for other institutions, as well as for the general reader who possesses a curiosity about matters botanical but a distaste for technical presentation. For a fairly simple and brief sketch of the plant kingdom, the reader should confine himself to a consecutive reading of the main text in the twelve "indispensable" chapters cited above. The sometimes rather generous footnotes which appear in these chapters, though not essential parts of the sketch, provide qualifications, amplifications, and speculations that are felt to be wholesome mental fodder for the student. For a somewhat more complete picture of the plant kingdom, the reader should read consecutively the entire twenty chapters. This complete reading is advisable for those students who plan to carry further their study of biology. It was the interests of this latter group particularly that were considered in the preparation of illustrations, which were drawn to resemble as closely as possible the actual laboratory specimens that such students would be likely to encounter in their subsequent course work.

The author is grateful to Dr. Paul Voth for his preparation of the illustrations, and to Dr. John Beal, Dr. Ralph Buchsbaum, Dr. George D. Fuller, Dr. Victor Johnson, Dr. George K. Link, Dr. Charles A. Shull, and Dr. Paul Voth for their very helpful criticism of portions of the manuscript.

MERLE C. COULTER

THE UNIVERSITY OF CHICAGO
July, 1935

TABLE OF CONTENTS

ix

CHAPTER I

PRIMITIVE PLANTS

THE botanist assures us that about 250,000 species, or distinct types, of living plants have already been discovered and described. Common experience tells us that within this vast kingdom there is tremendous variation in structural and functional characteristics. It appears to be a chaotic array of living forms until one finds some system which connects and relates all members of the plant kingdom in an orderly, understandable way. To date man has discovered only one fundamental concept which is successful in accomplishing this result—the concept of organic evolution. There is so much evidence to justify belief in this concept that biologists are convinced of its validity. Organic evolution derives all modern plants from one or a few very primitive ancestral types by a process of rather gradual modification through descent.

How these early ancestors were first brought into existence is another question, a question for which no one has as yet found a very satisfactory answer. Of course, biologists usually do their own speculating on this problem and come to favor certain hypotheses. The most plausible guess seems to be that the first living organisms upon the earth were derived from non-living materials already present; but just how this occurred remains a mystery, for man has not himself been able to produce life from non-living materials, or to discover anything of the sort occurring spontaneously in nature today.*

Granting that this first chapter in the story is (as yet) practically a blank, biologists maintain, however, that the subsequent chapters have been revealed, at least in their broad outlines. They feel com-

---

* Perhaps the different conditions of the remote past were responsible, and man in his few and possibly too brief experiments has not as yet properly duplicated these conditions. On the other hand, it may be that today life sometimes starts to originate as it did in the past but is regularly consumed before man can observe it by the ubiquitous established organisms of modern times.

paratively certain that the start was made by one or a few types of very simple organisms, existing at a period of a thousand million years ago or more. By a process of reproduction these early organisms perpetuated and multiplied their kind. For the most part the offspring resembled their parents, but occasionally one of them differed. This original difference may have been very slight, but, because it was perpetuated, it provided a starting-point for greater changes. Some generations later one of the descendants of the new type changed further, the second change being added to the first. Thus, by an accumulation of small changes over a great many generations, there emerged at last a type of organism that differed to a significant degree from its original ancestor. Meantime other lines of descent from this same ancestor may well have perpetuated themselves without modification. At the end of this period, therefore, individuals of the two types, old and new, existed side by side. The common ancestor had left two types of descendants. Evolutionary divergence had occurred.*

In our attempt to visualize the very primitive plants we might hope to get our most reliable clues from the fossil record. Actually that avenue of investigation has yielded very little information. The earliest plants apparently left little or nothing in the way of fossil

---

* Evolutionary divergence may follow one of several general patterns.

One pattern is characterized by a single divergence, i.e., a given ancestor leaves two lines of descent, one of which is unmodified, faithfully reproducing the characteristics of the ancestor itself, while the other line gradually becomes modified ("diverges") to culminate in descendants significantly different from the ancestral type. In such a case we would have among the organisms of today a living representative of the ancestor as well as the modified descendant.

Another common pattern is characterized by a double (or multiple) divergence. A given ancestor leaves two lines of descent that diverge in different directions, while the unmodified descendants are later exterminated. In such a case we would encounter today only the two modified descendants, but no living representatives of the ancestral type. Here the relationship between the two living forms might be more difficult to establish. If such a relationship were suspected, and a hypothesis to that effect were advanced, scientists would then seek a confirmation of the hypothesis. And if indeed the common ancestor had become extinct, the most convincing confirmation that one might hope to find would be the fossilized body of that ancestor, preserved in the sedimentary rocks of bygone ages. This would constitute the discovery of a "missing link."

remains; their tiny, delicate bodies did not lend themselves to fossilization, but were usually destroyed without a trace.*

With this line of inquiry blocked, we draw what conclusions we can from a survey of the forms that are still living. From a study of human history, and from our study of the later chapters of the fossil record, which are comparatively clear, we get the impression that things complex have usually been derived from things simple, rather than the reverse. By extending this idea we assure ourselves that the earliest plants were the simplest plants. We can do no better, therefore, than to give our attention to the simplest of the forms that are living today, on the assumption that these have descended with relatively little modification from the earliest ancestral types.

About the nearest approach to utter simplicity that we can find in living organisms appears in a small, rare, and decidedly inconspicuous group of plants known as the "blue-green algae," and of this group about the simplest of all is the genus *Gloeothece*.† The body of a single *Gloeothece* plant is far too small to be seen by the naked eye; we must examine it under the high power of the microscope to get any adequate idea of its structure. When we look at it in this way we see a tiny, nearly spherical body, consisting of a wall surrounding a mass of granular material on the inside. Actually, this material which looks granular is of a jelly-like consistency, and is nothing more or less than "protoplasm," the essential living substance, the truly living part of the body of every plant and animal. The surrounding wall, which is merely a lifeless product of the protoplasm itself, has a value in maintaining the shape and providing protection for the living substance within. The type of wall that we see here is a feature that serves fairly well to distinguish plants from animals. It is composed exclusively or in the main of "cellulose."

The protoplasm that we see in the body of *Gloeothece* should be

---

* In the earliest sedimentary rocks there are many deposits of carbon that suggest that living organisms were present, but do not tell us what they were like. In rare instances these rocks reveal the faint outlines of what may have been the bodies of simple algae.

† In the classification of plants and animals the term "genus" refers to the category that is next larger than that of "species." A discussion of such matters appears in chapter xx.

thought of not merely as a certain amount of the living substance but as being organized into a definite unit that we call the "cell." It is the organized protoplasm that constitutes the true cell, but in plants (not in animals) the protoplasm of the cell is usually surrounded by the cellulose cell wall.

Probably no concepts have been more significant and fruitful in the development of biology than those that are included in the so-called "cell principle." As it is usually employed the expression "cell principle" includes two component concepts: (1) that the bodies of all plants and animals are composed of cells and the products of cells;* (2) that new cells are derived only by the division of pre-existing cells.†

The higher plants and animals, such as ourselves, have multicellular bodies, with sometimes as many as several billion cells structurally and functionally co-ordinated in the body of a single individual. Many of the simpler plants and animals, however, have bodies that are unicellular. *Gloeothece* falls into this category, for the entire individual consists of only the one cell. This, then, is one reason for regarding *Gloeothece* as about our simplest plant; but this reason by itself would not suffice, for there are actually many thousands of types of plants which possess one-celled bodies.

Gloeothece is bluish-green in color, the effect of two liquid pigments, a blue and a green, which suffuse the protoplasm. The blue pigment is a comparative rarity in the plant kingdom, appearing only in the blue-green algae, and its function is not well understood. The green pigment, or "chlorophyll," however, is as famous as any substance in the biological world. Present in all green plants, it has the remarkable power of enabling the plant to manufacture food out of materials which themselves possess no food value. Green plants are, therefore, "independent" organisms, for they are capable of maintaining themselves in the absence of other forms of life. Here we see a second reason for regarding *Gloeothece* as primitive: the earliest organisms must have been independent. Once again, how-

---

* First enunciated for plants by Schleiden in 1838, and for animals by Schwann in 1839.

† Enunciated by Virchow in 1858.

ever, the criterion is not decisive, since this same independence is characteristic of most of the members of the plant kingdom.

In a state of nature, *Gloeothece* lives at the bottoms of shallow pools of fresh water. Some of this water diffuses through the cell wall into the protoplasm; also a certain amount of carbon dioxide, which is dissolved in the water. Out of these two simple raw materials, on the basis of the energy that is supplied by sunlight, and by virtue of its possession of the green chlorophyll, the protoplasm manufactures food for itself.

The food that is manufactured is utilized by the plant itself in three general ways.

Living protoplasm is a going concern, always in a dynamic state.* Like a running engine it demands a continuous supply of fuel. Otherwise it will stop running, and death will occur. For fuel the living organism can make use of only a limited class of substances—substances which not only contain energy but contain energy in a form that can be released and put to work by the organism. This is the category of substances that we refer to as food. *Gloeothece* manufactures its own food, and then consumes the larger part of it as fuel to keep its protoplasm alive.

Some food is stored up against a future need. If this were not the case the plant would rapidly die under those conditions (notably the lack of sunlight) which prohibited food manufacture. Among higher plants special storage depots are usually present. *Gloeothece*, however, can do no better than to store a certain amount of food rather diffusely through its protoplasm.†

A third portion of the manufactured food is devoted to growth.

---

* Actually there is much more motion within the protoplasm of actively living cells than is generally realized. This has been revealed in a sensational way in recent years through the expedient of "time-lapse photography." And in some plant cells the protoplasmic streaming is quite conspicuous even to ordinary microscopic observation. The "dynamic state" of living protoplasm, however, refers to far more than this matter of visible motion. The important and continuous chemical reactions and physical transformations that characterize living protoplasm are quite invisible.

† The idea of "food storage against a future need" should be sufficiently elastic to include the needs of the offspring as well as those of the individual

Displaying that power which, more strikingly than any other, distinguishes the living from the non-living, *Gloeothece* converts part of the food into additional protoplasm. Protoplasm, of course, is a very complex substance; its production involves not merely transformation of the food but also the addition of certain other chemical elements that are available in the surrounding medium. The resulting growth appears as an increase in the size of the *Gloeothece* cell, with a gradual stretching of its rather elastic wall and the production of new wall substance by the protoplasm.*

In the main features of food manufacture and food use, *Gloeothece* does no more nor less than any green plant. Its uniqueness lies in the fact that it accomplishes all this with a cell that is exceptionally simple. The protoplasm of *Gloeothece* is homogeneous; all parts of the protoplasm appear to be the same, and apparently all parts engage in all the life-activities. It is this simple, undifferentiated protoplasm that provides our third reason, and our best reason, for regarding *Gloeothece* as one of our simplest plants.

The fourth reason appears in connection with its reproduction. Under favorable conditions *Gloeothece* continues to grow rather steadily. When the cell has reached a certain size it simply pinches in two in the middle to form two small daughter-cells. These daugh-

---

itself. As we pass to the higher members of the plant kingdom we find increasing provision for food storage in general, and particularly for that which is to be of benefit to the offspring.

* To be more thorough we should assign this last fraction of the food not simply to "growth" but to "growth and repair." In many-celled organisms some cells of the body are frequently lost, either by accident or in the course of the ordinary life-processes. Animals have amazing powers and plants some powers of replacing these lost cells with new ones. Furthermore, within the individual cell itself, there are indications that older portions of the protoplasm are continuously breaking down and being replaced by new. This last is rather clearly revealed in animals by the nature of their waste products, and probably occurs less extensively in plants. Repair should be thought of as involving the same fundamental transformations of food as occur in connection with growth. Growth, however, brings an increase in the size of the body, which is not the case in repair. The phenomenon of repair should doubtless be admitted as a theoretical proposition for *Gloeothece* itself, but could hardly be demonstrated for such a simple plant as this.

ters round out into (nearly) spherical shape within what are now their own elastic cell walls.

This is the simplest conceivable type of cell division. At the same time it is an act of reproduction, for the plant body is single-celled. Two daughter-individuals* now exist where before there was but one parent-individual. Reproduction could be no simpler than this; so that here we have reason number four for regarding *Gloeothece* as the extreme of simplicity.†

The two new individuals proceed to carry out their lives quite independently of each other, and later reproduce according to the same simple program. Probably not much later, at that, for in such simple forms the "life-cycle" (i.e., the sequence of events that lies between a given stage in one generation and the corresponding stage

---

* The common use of the term "daughter" in such a situation is obviously incorrect, for sex is unknown in this simple organism. "Offspring" would, of course, be the better term, but actually "daughter" has crept into use in botanical parlance in situations where there can be no implication of sex.

† Botanists are in the habit of referring to the main body of any plant—the part that carries on the ordinary life-processes, or "vegetative" processes—as the "vegetative body." In the higher plants special reproductive organs are differentiated from the vegetative body, but in a blue-green alga, with its single-celled body, no such differentiation is possible. Here reproduction is accomplished by simple division of the vegetative body itself. Botanists express this idea when they say that the reproduction of *Gloeothece* is by "vegetative multiplication."

In this connection the point should be made that *Gloeothece* is primitive, not simply because it reproduces by vegetative multiplication but because this is the only mode of reproduction that it possesses. Actually many of the higher plants, while introducing new and more specialized methods of reproduction, retain as well the power of vegetative multiplication, so that at times they produce new individuals through separation from the parent-body of groups of what appear to be ordinary vegetative cells.

One might note wistfully that a one-celled organism like *Gloeothece* is "immortal." Death of the body from old age has no place in the reproductive program of such a form. The protoplasm goes on living forever in the bodies of the descendants (though, of course, some lines of descent frequently die off through starvation or "accident"). The higher, more complex organisms, such as man himself, have bought their specialization (and "breadth of experience") at the price of this immortality which their ultimate ancestors possessed (and retain only an "immortality" of a more restricted type).

in the next) is very brief. Under highly favorable environmental
conditions one generation in some of the blue-green algae may be
consummated in less than one hour's time. Perhaps we should cite
this feature, too, as a criterion of primitiveness for *Gloeothece*. Certainly it is a prevalent condition in simple organisms, while the more
complex bodies of higher forms must pass through quite a succession
of stages before they become mature and capable of reproducing.

One of the characteristics of blue-green algae as a group is the production of "mucilaginous sheaths." *Gloeothece* is no exception, for
apparently the outer part of its cell wall becomes changed (through

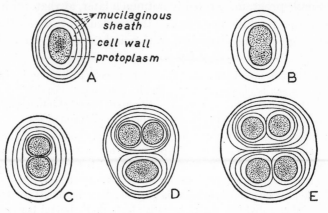

Fig. 1.—The blue-green alga, *Gloeothece*. *A*, single, one-celled individual; *B*, cell
division in process; *C–E*, colonies.

some action of the surrounding water) into a transparent zone of
mucilaginous material. Commonly two daughter-cells will remain
side by side, stuck in the matrix which is provided by the old mucilaginous sheath of the parent, and often the old sheath persists to
hold four "granddaughters" together. The two or four cells so associated are, however, mutually independent, so that we refer to such
a formation as a "colony" rather than a many-celled individual. If
any agency breaks up the colony, the individuals will apparently live
when separated quite as successfully as they had lived side by side
(Fig. 1).

Though the other blue-greens share with *Gloeothece* the several
other features of simplicity that we have described, the form of col-

ony produced by the others is somewhat more complex. Very often
thousands of individual cells are stuck together in a transparent
mucilaginous matrix which represents the combined output of all of
them. In nature, therefore, the blue-greens are most commonly encountered in the form of slimy masses (spheroid or amorphous),
growing in shallow fresh water or upon damp rocks. Some, however,
grow in salt water, and some on damp soil and the moist bark of
trees. One of the members of this group is responsible for the characteristic color of the Red Sea, showing that "blue-green" algae may
sometimes contain a conspicuous red pigment as well. They grow
successfully in hot springs at a temperature far beyond what most
other plants could endure. Around the hot springs and geysers of
Yellowstone National Park there are so-called "sinter" deposits, and
in connection with these there is a plentiful growth of blue-green
algae. Not only are these forms able to resist great heat but great
cold as well, great dryness, and strongly alkaline water. Altogether,
the resistance of this group exceeds that of all other plant groups,
save only the bacteria. The relation between their universality of
distribution and their high resistance is an obvious one. How these
two characteristics are related to the simplicity and antiquity of the
group provides an interesting field for speculation.

The closest relatives of blue-green algae are apparently the bacteria. In fact, it is this strong resemblance of bacteria to blue-green
algae that impels most biologists to place the former in the plant
kingdom. Aside from this, it is difficult to place the bacteria, for the
group displays a mixture of plant-like and animal-like characters.
Like the blue-green algae, the bacteria have single-celled bodies and
undifferentiated protoplasm; like the blue-greens, they reproduce
by simple cell division (vegetative multiplication) only, multiplying
very rapidly under favorable conditions, and like the blue-greens
many of them have extraordinary powers of resistance. In truth,
bacteria excel blue-greens in this respect, some being able to survive
boiling in water for several hours.

The big difference lies in the fact that bacteria lack chlorophyll
and cannot manufacture their own food. Hence they are "dependent" (most of them) directly or indirectly upon other living organisms. The combination of dependency, ubiquity, high resistance,

rapid multiplication, and microscopic (or even ultra-microscopic) size makes this group the great disease-producer for man and other organisms. For this and other reasons bacteria are of tremendous economic importance, and in recognition of this importance most universities now maintain distinct departments of bacteriology. We shall return to the bacteria in chapter xv, in a context that will bring out more clearly the significant rôles that they play in the organic world.

CHAPTER II

EARLY STEPS IN PLANT EVOLUTION

THE blue-green algae probably had some representatives living on this planet fully 1,000,000,000 years ago. At first there was doubtless just the one very simple type, but some of its descendants must have been modified as time rolled on, thus giving rise to new types, and in some instances perhaps "better" types. Today we recognize hundreds of existing species of blue-green algae, and there is no knowing how many additional species may be living their lives in such inconspicuous crannies as to have escaped the attention of the scientists. Furthermore, there may well have been thousands of species that were evolved in the past, only to be exterminated without leaving a trace.

Beyond these numerous lines of descent which changed only a little from the ancestral condition (so that all of them today have enough characteristics in common to be called blue-green algae), it is suspected that there was at least one line of descent which changed rather markedly from the others, and thus became the progenitor of a different group of plants. There is in existence today another group known as the "green algae." Though there is some difference of opinion as to the origin of green algae, their descent from the blue-greens appears most plausible to the writer.

Green algae, as the title implies, possess only the green pigment, chlorophyll, but they differ from the blue-greens in other characteristics as well. The most significant advance is in the differentiation of cell organs. Even such a form as *Pleurococcus*, which is about the simplest of the green algae, has taken this progressive step. No longer do we find the cell composed of homogeneous protoplasm; instead it is differentiated into three distinct parts which may be called "cell organs" or "organelles" (meaning "little organs"). In the center of the cell is a denser, spherical region of the protoplasm, sharply delimited from the rest by a definite membrane. This cell organ is known as the "nucleus." It carries the critical materials of heredity and exerts a governing influence over the rest of the cell. Between nucleus and cell wall is another dense and irregularly shaped cell organ known as the "chloroplast." In *Pleurococcus* the green chlorophyll, which was diffused throughout the entire cell of *Gloeothece*, is contained only within this chloroplast. Hence the chloroplast alone is endowed with the power to manufacture food. Surrounding both nucleus and chloroplast, and forming the more dilute matrix of protoplasm in which they lie, is the region of the cell known as "cytoplasm." It is the business of the cytoplasm to use the food manufactured by the chloroplast for maintenance and growth of the cell and to regulate the interchange of materials between the cell and the medium in which it lives. In short, the homogeneous protoplasm has

  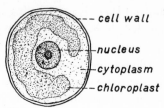

FIG. 2.—The green alga, *Pleurococcus*. *A*, single, one-celled individual; *B*, colony. Colonies are often very large and irregularly shaped.

FIG. 3.—Highly magnified section of *Pleurococcus*, to reveal relationship of cell organs.

been superseded by three specialists, each an expert at its own particular line of work (Figs. 2, 3). This type of change is usually regarded as a step in evolutionary progress.

We commonly think of progress as being any change in which the organism (individually or as a member of some group) achieves a more efficient adaptation to the conditions of life. This result may be brought about in any one of a great many ways. *One* of the commonest ways is by a change from the relatively "generalized" to the more "specialized" condition. History assures us that progress in human societies has frequently involved this very thing. In the primitive social group—as, for example, in some pioneer community—all those composing the group are likely to be doing pretty much the same work. As time passes, some usually come to devote themselves to some tasks, and others to others. By this concentration of attention upon one task, this freedom of responsibility from all other duties, a man may become quite expert. The social group as a whole, by co-ordinating the work of its several experts or specialists, will thus get the sum total of its tasks performed more efficiently than before. Increased efficiency means a better adjustment to the requirements of life. It makes possible a discharge of necessary duties with comparative ease, leaving a comfortable margin of available strength which may be called upon in times of stress.

Such considerations make us regard transitions from generalization to specialization as steps in progress (though there may be exceptions to this principle in special situations). Accordingly, *Gloeothece*, the blue-green alga, with its generalized protoplasm, appears to be quite primitive, while *Pleurococcus*, the green alga, with its differentiation of cell organs, appears to have taken a step in evolutionary progress.

Of the thousands of known species of green algae, all possess this same differentiation of cell organs. Size and shape of cell vary tremendously among the different species, as do also the size, shape, and number of chloroplasts, and other characteristics as well. Yet all species retain the same fundamental organization of the cell into nucleus, chloroplast, and cytoplasm, and this makes us believe that they have all inherited this organization from a common ancestor. Was this common ancestor *Pleurococcus?* Clearly the *Pleurococcus* individuals that are alive today cannot be the ancestors of contemporary species. To be sensible the question can only mean, "Was the common ancestor of all other green algae so closely similar to the living *Pleurococcus* that it would be classed as a member of the same genus?" This question, like the innumerable similar questions that are asked in biology, cannot be answered with any assurance. The best that one can ever say is something like this: "Probably all of the green algae have been derived from a single progenitor, and probably this progenitor resembled *Pleurococcus* in having a body composed of a single, spherical cell, in showing a simple form of the differentiation of cell organs, and in possessing only a very simple method of reproduction."

In at least one respect *Pleurococcus* differs from the ideal, hypothetical ancestor of green algae. One would infer that this ancestor grew submerged in fresh water, for practically all of the modern green algae (like most of the blue-greens) are to be found in such places. *Pleurococcus*, on the other hand, most commonly grows not actually submerged in water but in places that are moist or usually moist. One recalls seeing a thin green coating over parts of the surfaces of bricks, stones, and flower-pots that have been kept for a long time in a moist greenhouse. One may also recall a similar green coating on the bark near the base of trees in a moist forest.* These coats commonly consist of nothing but innumerable microscopic bodies of *Pleurococcus* crowded together. Of course, a few raindrops or a minute film of dew will effectively submerge these plants and enable them to carry on their life-processes in the manner that is customary for their group. But this submergence occurs only at intervals. *Pleurococcus* is unique among green algae in being able to survive the intervening periods of comparative drouth, due perhaps to some special quality of the cell wall or the protoplasm itself.

A *Pleurococcus* cell, which is at the same time a *Pleurococcus* individual, grows to a certain size and then divides.† In this case divi-

---

* In this part of the world *Pleurococcus* is usually to be found more abundantly on the north than on the south sides of trees. Lost in woods, and without a compass or the sun to guide him, one might refer to these *Pleurococcus* growths as direction-markers.

In the northern hemisphere some of the direct rays of the sun strike the south but not the north sides of the trees. One might jump to the conclusion that the increased heating and evaporation suppresses the growth of *Pleurococcus* on the southern exposure. Critical information, however, assigns the effect to the greater dosage of ultra-violet rays, which are destructive to the protoplasm of those organisms which lack adequate protective devices.

† The size attained by the cell at the moment of division varies somewhat— varies more, in fact, than in some other unicellular forms that might be cited— and yet one is impressed with the fact that *Pleurococcus* cells are all pretty much the same size at the time they divide to form two daughters. Not that there is anything unique about this, for one is equally impressed by the corresponding phenomenon in many other organisms. Why these fairly uniform size limits?

If we can answer this question for a one-celled organism, we may thereby be suggesting at least one part of the answer for such a complex organism as man. Biologists have usually guessed that the answer is to be found along one or both of the two following lines:

1. Normal life-activity depends upon the diffusion of materials in (and out)

sion is not accomplished by the simple process of pinching in two that is employed in blue-green algae. The protoplasm of *Pleurococcus*, with its several organelles, is a heterogeneous system, such that any rather crude bisection of the cell might often yield two daughter-cells with significantly different contents. Since actually the characteristics of *Pleurococcus* appear to be rather perfectly perpetuated from one cell generation to the next, the organism must be endowed with some more accurate method of division. As a matter of fact, the cells of *Pleurococcus*, like practically all of the cells of higher plants and animals, divide by the process that we call "mitosis." Mitosis involves a series of events that bring about a remarkably ac-

---

through the cell surface to (and from) the mass of protoplasm. The more the surface, the more the *possibility* of exchange of materials; the more the mass of the protoplasm, the more the *need* for exchange of materials. Double the diameter of the cell, and you square its surface. But, at the same time, you cube its mass. Hence, with increase of size, surface keeps diminishing relative to mass, and a ratio is finally reached where the cell is presented with the alternative of continuing to grow and reducing its life-activities, or dividing and maintaining the life-activities at approximately their normal rate.

2. The nucleus "governs" the cell as a whole. Growth of a cell is growth of the general cytoplasm (and chloroplast) only, the nucleus maintaining essentially its original size. When the "domain" reaches a certain size, effective government requires two governors instead of one.

These two suggestions concerning the division of the cell at the time that a certain size limit has been reached should not be regarded as genuine explanations, for they tell us nothing of the causal mechanics involved, of the actual physical or chemical forces that are responsible for initiating the division. Instead they merely point out how a division at this stage is advantageous to the living cell.

In the study of biology one is repeatedly encountering these two types of interpretation of the structures and functions of living organisms: (*a*) statements as to how the structure or function may be advantageous to the organism: (*b*) descriptions as to how materials and energy have been applied to the production of the structure or the discharge of the function. In the interests of clear thinking the students should form the habit of distinguishing these two types of interpretation, and of regarding only type (*b*) as constituting a genuine explanation. With a view to the clearest possible understanding of biological phenomena, biologists are constantly searching for such genuine explanations. In many cases, however, where the genuine explanation has not yet been reached, the biologist feels that he has at least advanced a little in his understanding of the phenomenon by providing an interpretation of the (*a*) type.

curate division of the contents of the nucleus, so that each daughter cell receives a nucleus qualitatively like that of the mother-cell. Apparently it is the nuclear contents that are of most critical importance in deciding the characteristics of the cell. While the nucleus is being accurately divided by mitosis, chloroplast and cytoplasm are being pinched in two in a less accurate way, and a wall comes in to separate the two resulting daughter-cells.*

These daughters usually remain attached to each other along the line of division, and, since such attachments may persist through many cell divisions, a single cell may at last yield an irregular clump that is composed of scores or hundreds of descendants. Perhaps this attachment is due to the facts that *Pleurococcus* lives in moist air and has no power to move its body. Most of the one-celled green algae live in water and are endowed with means of locomotion, so that after division the two daughter-cells become separated. Even in *Pleurococcus*, however, the cells appear to be functionally distinct individuals, for, if they chance to become separated, they continue to carry out their life-processes in the same way as they did when they were attached.

In summary, the contribution that *Pleurococcus* makes to the evolution of the plant kingdom lies in the differentiation of organelles and a new method of cell division that is associated with that differentiation. Most of the green algae, however, have taken still further steps in evolutionary progress.

On the stones that lie in shallow, fresh, rapidly-moving water one sometimes encounters green growths that have a rough resemblance to human hair. Such growths may consist of any one of a number of types of green algae, one of them being the genus known as *Ulothrix*. Under the microscope *Ulothrix* is revealed to be a long filament, composed of a single line of cells attached end to end. When growing conditions are good, each cell contains a nucleus, cytoplasm, and a single large chloroplast (shaped like a napkin ring, and surrounding the bulk of the cytoplasm and the nucleus (Fig. 4*A*).

All cells of the filament are essentially the same, with two conspicuous exceptions: (1) the cell at the upper or free end is dome-

---

* The student would do well to defer his inquiry into the exact nature of mitosis until a later stage in his biological training.

shaped rather than cylindrical. A living cell normally takes in enough water to exert a strong outward pressure upon its retaining wall. It follows that when the wall is elastic (as is commonly true of the cellulose cell walls of plants), and equally elastic in all places, the cell will assume the shape of a perfect sphere. In *Ulothrix* mutual pressure of adjoining cells flattens out their end walls and converts the sphere into the barrel shape. The terminal cell, however, being unconfined above, becomes a dome. (2) The lowest cell of the fila-

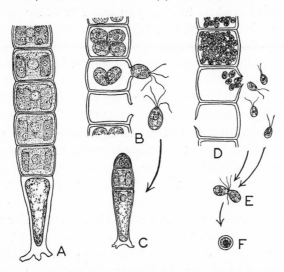

FIG. 4.—*Ulothrix. A*, vegetative filament with holdfast cell; *B*, production and discharge of spores; *C*, young filament recently produced by a spore; *D*, production and discharge of gametes; *E*, fusion of gametes; *F*, zygote.

ment, known as the "holdfast cell," is longer and narrower than the others, and its lower surface, conforming to irregularities on the surface of some submerged rock or other object, adheres thereto, and thus serves as an anchor for the entire filament. By virtue of the holdfast, *Ulothrix* is able to maintain itself in a medium of rapidly moving water. Through evolving forms equipped with holdfasts the green algae have been able to exploit a range of "habitats" that would otherwise have been unavailable.*

* Habitat refers to a particular complex of environmental conditions. Accustomed as we are to the versatility of man, it surprises us to find that most lower

Food manufacture by the individual cells and their resulting growth leads to cell division. In *Ulothrix* cell division is restricted to a single plane, and it is this which maintains the filamentous form of the entire body. Simple cell division of this sort does not effect reproduction of the entire body but merely increases its length.

The plant body is reproduced by a somewhat more involved program. Under certain conditions the entire protoplasm of a cell will shrink slightly away from the cell wall and divide several times. The result is anywhere from two to thirty-two tiny protoplasmic bodies, each being actually a small cell equipped with nucleus, chloroplast, and cytoplasm, but all contained within the confines of the old cell wall. After a time the cell wall breaks and these bodies are released into the water. Each one is pear-shaped and equipped at the more pointed end with four long and extremely thin extensions of the protoplasm that are called "cilia." At this stage the protoplasm is relatively "naked," lacking the cellulose wall which ordinarily surrounds plant cells, and being held in shape merely by the "protoplasmic membrane," i.e., the stronger and less fluid outermost layer of the protoplasm itself (Fig. 4*B*). All cells of the filament commonly produce and discharge these bodies at about the same time, and the many filaments in the same vicinity act in the same way almost simultaneously.

The bodies that have been discharged are known as "spores," our first example of a type of reproductive unit that appears in all plant groups but a few of the lowest. The spores of *Ulothrix* swim about very actively for a time, being propelled by the lashing action of their cilia. At the same time, of course, they are usually being carried along in currents of water. At last they quiet down, their cilia rapidly disappearing, and a small few of the multitude of spores may chance to be lodged in situations that will permit them to produce new plants. The successful spore, still a naked bit of protoplasm, fits itself like a tiny drop of jelly into irregularities in some sub-

organisms can succeed in only one habitat (or a very few). The earth's surface, of course, provides this particular habitat at many points. All such points will be occupied by the species in question, provided (1) that the species has ever been distributed to that spot, and (2) the spot has not been effectively preempted by other forms of life.

stratum at the bottom of the water, and by this act establishes the holdfast cell that is to be.

The spore is now on its way to becoming a young plant. Equipped with a chloroplast, it commences food manufacture and has soon grown to the size which evokes cell division. From here on it is merely a story of continued growth and cell division until a full-sized filament results (Fig. 4C). By means of these spores *Ulothrix* may run through quite a number of generations in a single growing season.

It appears, then, that the spore can accomplish reproduction of the many-celled body where ordinary cell division fails to do so. But, beyond the matter of reproduction itself, or the simple bringing into existence of new individual bodies, there are two other functions that spores discharge.

The ultimate success of any species depends not merely upon how well its body may be adapted to the environment but also upon its reproductive efficiency. Of the many elements which enter into reproductive efficiency, a very obvious one is the reproductive ratio. In simple organisms, such as *Gloeothece* and *Pleurococcus*, the reproductive ratio is two, meaning that one parent-individual gives rise to two offspring. Other things being equal, the more successful species will be the one with the higher reproductive ratio. The spore reproduction of *Ulothrix*, in which the scores or hundreds of cells of the many-celled body may each yield several spores, and each spore has the power of producing an entire many-celled body, provides a very high *potential* reproductive ratio. The *actual* reproductive ratio will depend, of course, upon how many of the spores succeed.

Another function of the spore is that of distributing the species. Those species which have made the most successful conquest of the earth's surface are those which have had effective means of scattering themselves to new locations. In many plants the spore is practically the only agency that can bring about this result. The better to insure it, spores of *Ulothrix* and of most other water forms are equipped with cilia which scatter them through the medium even in the absence of water currents.

Spores, however, are not the only means of reproduction in *Ulothrix*. Conditions which exist toward the end of the normal growing season are apparently responsible for stimulating the vegetative cells to a new activity. As in spore formation, the protoplasm withdraws from the wall and starts to divide. In this case, however, a greater number of divisions occurs, so that the final product is likely to be sixty-four or more tiny protoplasmic bodies within the old cell wall. Here, too, each of these products is a complete cell, with its nucleus, cytoplasm, and a tiny bit of chloroplast. As before, these bodies are discharged with the breaking of the cell wall, as before they are equipped with cilia (two, in this case), and as before they swim around quite aimlessly for a time (Fig. 4D). It appears, however, that these tiny bodies are incapable of producing new filaments by their own independent action. Only through another type of action can they be instrumental in bringing about reproduction. With such a great swarm of these bodies swimming at random in the same small region it is inevitable that occasionally two will collide. Though many of these collisions come to naught, there are some that bring a startling result, the two tiny cells fusing together to form one (Fig. 4E).

This is clearly a sex act, here revealed in what is probably its most primitive form. The fusing bodies are called "gametes" and the fusion product a "zygote."

The cilia of the gametes have disappeared in connection with the fusion, so that the zygote soon settles down and lodges at the bottom of the water. Surrounding itself with a heavy, protective wall, the zygote passes into a period of dormancy (Fig. 4F).* In this condi-

---

* Dormancy is sometimes carelessly referred to as a state of "suspended animation." The implication, however, is inconsistent with modern biological thought. If by animation we mean the carrying-on of fundamental life-processes in the protoplasm, then to suspend animation, even briefly, brings death. Dormancy is not death, but merely a condition in which the life-processes are being carried on at an extremely low rate, even much lower than in the deepest sleep.

Though it is very common among primitive plants for the zygote to pass at once into this dormancy, and thus provide a "wintering-over" stage for the species, it should not be concluded that this behavior represents the true function of sex, or even that it has any *essential* connection with sexual reproduction. Almost any simple organism will respond to hard conditions, at almost any stage of the life-cycle, by the production of heavy-walled, dormant cells. When we recall that *Ulothrix* gametes are usually produced toward the end of

tion it usually lasts over winter, to be awakened by the more favorable conditions which arrive in spring. At that time the zygote divides internally to form four spores, which escape through the broken zygote wall and are capable of producing new filaments by their own independent action.*

Practically all of the green algae include some form of sex among their methods of reproduction. Many of them are like *Ulothrix* in having similar gametes, but more of them have taken a further evolutionary step and possess gametes of two distinct types, male and female. This may be illustrated by a fairly common form known as *Oedogonium*.

---

the growing season, we realize that the newly formed zygote will, in all probability, soon be surrounded by those conditions which provoke dormancy. It is, therefore, the conditions which obtain at the time that best explain the dormancy of the zygote, rather than the mere fact that it is the product of a sexual union.

* A consideration of such forms as *Ulothrix*, in which the gametes are similar (rather than differentiated into male and female gametes), comparatively simple, and apparently little more than small editions of the spores, has suggested that sexual reproduction originated from spore reproduction. It would appear that the first sex acts ever to occur were nothing more than purely accidental fusions on the part of undersized spores. There is nothing unprecedented in such a view, since most biologists trace back the origin of all the structures and modes of behavior of higher organisms to "happy accidents" in the first instance.

The accident which brings a benefit to the species is usually perpetuated by inheritance. So we must conclude that some benefit resulted from sex. An exact understanding of the function of sex demands a knowledge of the machinery of heredity. Even without such knowledge, however, one can catch the general import of the following considerations. For evolution to occur there must be variation in the population, i.e., some individuals must possess hereditary characteristics different from the rest. The amount or rate of evolution will depend, among other things, upon the amount of variation that occurs. Sex is conducive to a greater amount of variation than would otherwise occur, since sex is the only form of reproduction that can combine hereditary characteristics from two different parents, and thus open the way to a series of new combinations of characters in the individuals of the following generations. Thus the advantage of sex is in encouraging evolution and in this way favoring the production of species that are better adapted to their environments. One can see from this that the real benefit of sex is realized only when the two gametes come

Oedogonium, like the majority of green algae, has a simple filamentous body that is anchored to the substratum by a holdfast cell,
but is usually found growing in quieter water than is *Ulothrix*. The
vegetative cell is longer and the chloroplast of a more irregular shape,
but the more significant differences lie in the reproductive structures. Under certain conditions one or more cells of the filament will
each discharge a single large spore, pear-shaped, and equipped near
the more pointed end with many cilia in crown-like arrangement. In
its behavior, however, this swimming spore is like the smaller and
structurally simpler spores that we saw in *Ulothrix*.

It is in connection with the gametes that we see a pronounced
change. Under conditions favoring gamete production, an occasional
vegetative cell in the filament will enlarge, bulging outward and becoming more spherical than its neighbors. This is no more than the
superficial consequence of a transfer of some of the nutritive materials from neighboring cells to the cell in question. The result is a
very large gamete, well filled with food, but with no power of motility. This large passive gamete is the female gamete, or "egg" (Fig.
5*A*).

Elsewhere in the same filament male gametes are being formed at
the same time.* In this case a vegetative cell will divide into a series
of smaller cells separated by walls. Each of these smaller cells then
divides internally to form two small, active gametes, the male
gametes or "sperms" (Fig. 5*B*). When the retaining walls break, the
sperms, structurally small replicas of the spores of the same species,
swim actively and, for a time, at random. Some of them, however,
are soon directed in their swimming by a chemical that is exuded

---

from parents of a somewhat different hereditary constitution. Even in *Ulothrix*
this is at least a possibility, since many *Ulothrix* filaments are ordinarily releasing
gametes into the water at the same time; and one recent investigator claims to
have demonstrated that the *only* successful fusions are those of gametes from
different filaments. (An account of the origin of the *original* hereditary differences by the process of "mutation" is presented in chap. xix.)

* In some species of the genus *Oedogonium* some of the filaments produce eggs
only, while others produce sperms only. In still other species full-sized filaments
produce eggs only, while the sperms are produced by "dwarf male" filaments
that become attached, with their tiny holdfasts, to the larger filaments.

from a neighboring egg. As a result of this guidance, they swim toward a small aperture in the cell wall which retains the egg, and through this the first sperm to arrive makes its entry. The ensuing fusion of sperm and egg, often referred to as the act of fertilization, produces a zygote. Almost instantaneously the protoplasmic membrane (not the wall) around the egg changes in such manner as to prohibit the entrance of any sperms that might arrive subsequently.

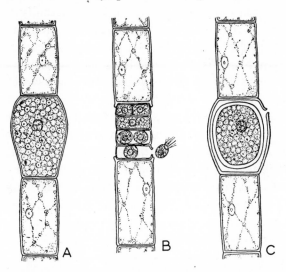

Fig. 5.—*Oedogonium*. *A*, portion of filament containing unfertilized egg; *B*, production and discharge of sperms; *C*, a zygote with its newly formed, heavy wall; note aperture in original wall through which sperm entered.

The *Oedogonium* filament, which now carries, here and there along its length, several fertilized eggs or zygotes, is destined soon to disintegrate, releasing the zygotes from their previous confinement. By this time, however, each zygote has laid down around itself the heavy wall which is to provide protection during the ensuing period of dormancy (Fig. 5C). The rest of the story duplicates that of *Ulothrix*.

Ulothrix possessed similar or undifferentiated gametes. With the differentiation of gametes *Oedogonium* has made a significant evolutionary advance, for this same distinction between large passive egg and small active sperm is preserved among all higher forms. The biological value (i.e., value to the species) of differentiation of gametes is readily seen. Whether or not an individual is to survive depends in large part upon what sort of a start it makes, for it is the period of infancy that is the most defenseless, the most susceptible to adverse environmental influences. *Oedogonium*, through the food that is stored in the egg, furnishes its young with a greater nutritive capital than does *Ulothrix*. In this way they are, in effect, "fed by the parent" until they have had ample opportunity to establish themselves. So the differentiation of gametes is an improved provision for the young. Further steps in this same general direction are taken by still higher plants, while the same evolutionary trend appears even more impressively as we pass from the lower to the higher members of the animal kingdom.*

With the differentiation of gametes there is established a reproductive program in which it might appear that the maternal parent contributes a great deal more to the offspring than does the paternal parent. This is true in only a limited sense. The female parent does indeed contribute more nutrition than does the male, and nutrition is highly important in deciding whether the young will pull through the early critical period of life. But this nutrition has no more effect in deciding the specific and varietal characteristics of the organism

---

* Earlier it was stated that the long-run success of a species depended upon its reproductive ratio. Obviously it must be the *effective* reproductive ratio that has this significance, and not merely the *potential* reproductive ratio. Equally obviously the effective reproductive ratio will depend upon the total number of reproductive units produced and the proportion of those units which come through successfully in the establishment of individuals of the new generation. Organisms achieve large effective reproductive ratios in either of two general ways. Some give their young a rather poor start, but produce such a prodigiously high number of reproductive units that quite a few young are likely to be established in spite of a high mortality. Others produce fewer reproductive units but give their young such a good start in life that the mortality is much lower. Though more than a few exceptions might be found to the generalization, it does appear to be generally true that lower plants and animals emphasize the high potential reproductive ratios, while the higher types (particularly the higher animals) have the lower potential reproductive ratios but make more provision to give their young a good start.

than does the soil in which a plant grows. The physical basis for the truly hereditary characters lies in the nucleus. In any case of differentiation of gametes the difference is in the cytoplasm, while the nuclei of egg and sperm are of the same magnitude and importance. Experiments in heredity have amply demonstrated that (with negligible exceptions) male and female parents contribute equally to the hereditary constitution of their young.

CHAPTER III

THE HIGHER ALGAE

IN HIS classification of plants the botanist starts by dividing up the entire plant kingdom into four great groups. Of these four groups the one that contains the simplest plants goes by the name of Thallophytes.* Thallophytes, in turn, are usually divided into two or three subdivisions. One large subdivision is made up of the green, independent algae; another of the non-green, dependent fungi. Some botanists assign all Thallophytes to these two groups, while others borrow members from both of these groups to make a third (and simplest) group, containing the blue-green algae and the bacteria. Algae, in turn, are divided up into four (or three) groups, the blue-green algae (Cyanophyceae), the green algae (Chlorophyceae), the brown algae (Phaeophyceae), and the red algae (Rhodophyceae).

To date we have discussed only the blue-greens and three members of the green algae. Before looking at any additional forms we would do well to pause and consider the matter of the evolutionary transition from one-celled to many-celled bodies. Mere increase in size of an organism (or of an institution) is not a truly progressive step. Progress results only in so far as the many-celled body can accomplish certain things that the one-celled body fails to accomplish,

---

* Or, more technically, *Thallophyta*. The last part of the word comes from the Greek "phyton," meaning "plant," and reappears in titles of the three other great groups. The first part is derived from "thallus," which signifies a plant body in which there is little or no differentiation of vegetative parts. Actually the title is not quite definitive, since a few members of this group show a greater differentiation of body regions than do a few of the members of the next higher group. A more serviceable distinction is based on the condition of the "sex organs." As in *Oedogonium*, the gamete-containing structures of Thallophytes are one-celled, containing the gametes within the cell-wall only. As we shall see later, the sex organs of the next highest group are many-celled affairs, the gametes being covered over by a complete layer of protective cells.

26

or can discharge certain functions more efficiently. To a limited extent this may be true even when all the cells of the many-celled body are alike. (We shall consider an example of this in the chapter on the acquisition of the land habit.) For the most part, however, the advantage of the many-celled condition lies in the fact that it opens up the possibility of specialization on a new level. If one cell, or preferably a whole "tissue" of cells, is specialized to discharge a certain function (and if many such cells or cell groups are co-ordinated in their activity), then the body as a whole may constitute a more efficient working unit than does the one-celled body.

The higher algae must have become many-celled ages ago, but apparently they never made much of the possibilities of the many-celled body.* Some advantage is derived from a specialized holdfast at the bottom of the filamentous body, for in this way algae can maintain themselves in rapidly moving water. Thus the group as a whole can exploit quite a range of habitats that would otherwise be unavailable. Some advantage is gained in combining food that is derived from several cells into the large egg of *Oedogonium*, for in this way the young are given a better start in life. Some advantages, too, are derived from the somewhat more complex body organization that we shall see among brown algae. It was not until much later in evolutionary history, however—not until life had emerged from the water and started to spread over the land surface—that plants really capitalized upon the possibilities of the many-celled body.

Man, who is forever classifying things, finds it convenient to put living organisms into the two great categories of one-celled individuals and many-celled individuals. No doubt such a treatment is convenient, but at the same time it is in danger of being misleading. It seems to suggest that the delimitation of the two categories is sharp, corresponding to a sudden, radical evolutionary change that occurred in the past; and this is far from the truth. If we look, on the one hand, at the simple one-celled body of *Gloeothece*, and, on the other hand, at a human being with his several billion cells, highly interdependent but co-ordinated into the single functional unit which

---

* Perhaps this was because their environment and mode of life were so simple that there was nothing to evoke a greater complexity.

provides our best example of "individuality," we may well be impressed with the profound difference between one-celled and many-celled organisms. These, however, are the extremes, and if we give our attention to some of the intermediate cases we find great difficulty in determining just where the condition of the one-celled individual leaves off and the condition of the many-celled individual begins. For actually there is no sharp delimitation of the two great categories in nature, and such evolutionary change along these lines as has occurred in the past must have occurred very gradually indeed.

There is no plant in existence that possesses anything like the full development of individuality that appears in man or in any of the higher animals. At their very best, many-celled plants are but feebly individualized. Even so, such a thing as an oak tree can, with justification, be referred to as a many-celled individual, for its numerous cells are living an interdependent life and each one is playing a rôle in the economy of the larger organization. Between this condition and that of *Gloeothece* there are many transitional levels.

The mere fact that the cells of *Gloeothece* or the cells of *Pleurococcus* are usually found growing in groups is not enough to qualify them in our minds as many-celled individuals. The group does not act as a unit, and there is no evidence of interdependence on the part of the component cells. We say, therefore, that such a group is merely a "colony" of single-celled individuals. Then what of *Ulothrix* and of *Oedogonium?* In the botanical literature these forms are referred to in both ways, both as colonies of single-celled individuals and as many-celled individuals. The writer prefers to think of them as colonies of single-celled individuals that have taken a few small steps in the direction of the many-celled individual, for such a diagnosis helps to keep alive the undoubted fact that in nature the transition has been a gradual one. The higher algae have

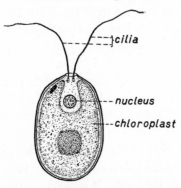

Fig. 6.—*Chlamydomonas*, a single-celled, motile green alga.

taken further steps in the same general direction, but there is not a single alga and not a single Thallophyte that has achieved any very convincing many-celled individuality.

The green algae made little evolutionary progress with respect to the complexity of the vegetative body. In some the body is merely a single cell, stationary, as in *Pleurococcus,* or more often free-moving in the water, either passively or actively by means of its cilia (Fig. 6).* Most commonly, however, the body is a filament, one-celled in

---

* Biologists have always been puzzled by a group of one-celled organisms which are usually referred to as "flagellates." The name refers to the "flagellum," a long and usually single cilium-like structure, by means of which these organisms propel themselves through their liquid medium. The flagellates lack the cellulose cell wall that is so characteristic of plants, and are surrounded by nothing but a (slightly stiffened) protoplasmic membrane. In consequence the jelly-like body can assume a great variety of shapes. Some flagellates are endowed with chloroplasts, by means of which they manufacture a certain amount of their food, but supplement this by the "ingestion" of solid food particles in typical animal-like fashion (Fig. 7). There are others which, though structurally quite similar, lack the chloroplasts and live strictly dependent lives. A number of these are great disease-producers. One, for example, sometimes invades man's blood stream to produce the well-known "sleeping-sickness" of Africa.

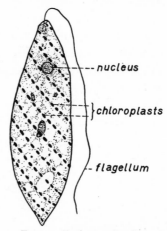

FIG. 7.—*Euglena,* a flagellate

In the flagellates, therefore, we have a group that is difficult to place in either the plant or animal kingdom. Some biologists, in fact, put them in a kingdom of their own, the "Protista," or "primitive plant-animals," from which they say that both plant and animal kingdoms have been derived.

Volvocales is an order that is usually assigned by the botanist to the green algae. The group is characterized by vegetative cells which subsist only on the food which they manufacture by their chloroplasts, and which are endowed with (usually two) cilia. Some members of the group exist singly and others form smaller or larger colonies. In the famous *Volvox,* for example, a colony

caliber. This filament may be unbranched, as in *Ulothrix* and *Oedogonium*, or may be divided into a series of filamentous branches (Fig. 9). A very few have plate-like bodies, in which a single layer (sometimes a double layer) of cells is spread out in two dimensions (Fig. 10).*

In one group of algae (the Siphonales) we find the "coenocytic" condition. Here the body is a branched filament, but there are no

---

may consist of thousands of individuals arranged in the form of a hollow sphere (Fig. 8). Reproductive methods, in some forms reminiscent of *Ulothrix* and in

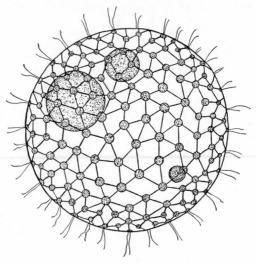

FIG. 8.—A *Volvox* colony, with its ciliated vegetative cells arranged in the form of a hollow sphere. Some of the cells have the power to enlarge and divide to form new colonies; three such are shown within the interior of the sphere. The eggs and sperms are not shown in this particular specimen.

others of *Oedogonium*, constitute an additional plant affinity. Yet the Volvocales grade into the flagellates, and along with the flagellates are assigned by some zoölogists to the animal kingdom, and by others to the Protista.

* In one group, the Charales or stoneworts, the body is a branched filament several cells in caliber. Though the Charales are usually treated as a subdivision of the green algae, it is recognized that the complexity of their bodies, and particularly of their reproductive structures, puts them in a distinct class by themselves, some botanists claiming that they should not even be included among Thallophytes.

cross walls, the filament containing a continuum of cytoplasm, spotted here and there with nuclei (Fig. 11).*

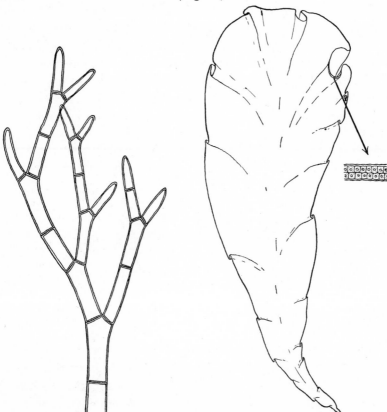

FIG. 9.—Outline sketch of the common green alga, *Cladophora*, as an example of the branching filament.

FIG. 10.—Sketch of *Ulva*, the "sea lettuce," rather commonly found washed up on the New England coast. The enlarged fragment, shown in section, reveals the two-layered condition of the body.

---

* In such a case the "cell" is not structurally delimited. In the functional sense it might be assumed that one cell in a coenocyte is represented by one nucleus plus the adjoining cytoplasm which constitutes its domain, but such an assumption would serve to clarify no biological problems. Some biologists state that, while most organism. have their bodies organized on a "cellular" basis, there is a "non-cellular organization in the coenocyte, where any theoretical individuality of smaller units is lost in the organization of the body as a whole.

The bulk of green algae grow submerged in fresh water, most of them anchored to the bottom by their holdfasts, but some (e.g., *Spirogyra*) floating free at or near the surface of stagnant waters. A few are found in the ocean (e.g., *Ulva* or "sea lettuce"), and some in air which is continuously or frequently very humid, as on a moist soil surface or the bark of old forest trees (e.g., *Pleurococcus*).

In reproductive devices the group made more progress. Already we have noted that the green algae added to the vegetative multiplication, which they must have inherited from their ancestors, the

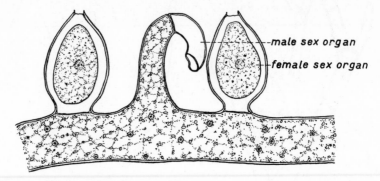

FIG. 11.—Fragment of the green alga, *Vaucheria*, showing coenocytic body and differentiated sex organs.

new methods of spore reproduction and sexual reproduction which were to mean so much to their descendants. In addition they introduced the differentiation of gametes, a further step in the advancement of sexual reproduction. Beyond this two more refinements in the program of sexual reproduction are added by some of the higher green algae.

The term "sex organ" is applied to any container of gametes.* In a form like *Oedogonium* the sex organs appear at random along the filament, being nothing but transformed vegetative cells. In such a form as *Vaucheria* (a coenocyte), however, we see that special side branches are set apart and destined from the beginning of their ex-

---

* Among Thallophytes the simple structure which contains the egg is technically referred to as the "oögonium" (equivalent to "female sex organ"), and that which contains the sperms as the "antheridium" (equivalent to "male sex organ").

istence to become sex organs (Fig. 11). Here, then, we note a "differentiation of sex organs," which might well have been predicted from
the observation that "generalization is usually succeeded by specialization."

The other refinement consists of the "differentiation of sexual individuals." As has been pointed out before, sex brings an advantage
to the species only when the two gametes have been derived from
bodies which differ somewhat in their hereditary characteristics. It
is not surprising, therefore, to find that nature is full of devices which
favor and in some cases insure "cross-fertilization." Most obvious of
all devices to insure cross-fertilization is that of a sexual differentiation of individuals. If the one plant can produce only sperms and
another plant can produce only eggs, any fertilization that occurs
will be cross-fertilization.

Best known of all the green algae is the picturesque *Spirogyra*—
picturesque, not for its habitat, which is stagnant water, but for its
unique chloroplasts, which are coiled like green ribbons within the
walls of the cylindrical cells that make up the filament. *Spirogyra*
produces no spores, but relies upon sex for its reproduction. Among
the commoner species of this genus, male and female filaments exist.
Indistinguishable in the vegetative state, the two filaments reveal
their sexes when they chance to float side by side. Under those circumstances tubes grow out from the cells of each filament—tubes
which join at their tips so that the two filaments are now connected
all along their length in a ladder-like arrangement. After a time the
adjoining end-walls of the tubes are dissolved away, creating passageways which lead from each cell of one filament to the neighboring cell of the other. At this stage the entire protoplasm of each cell
—spiral chloroplasts and all—pulls away from its wall and rounds
up. The two sexes cannot be distinguished by size but only by their
activity. All of the rounded protoplasts of the female filament remain quiescent, while those of the male filament force themselves
through the narrow connecting tubes to fuse with the female gametes on the other side (Fig. 12).*

---

* Though the program described is the one most commonly encountered in
this genus, there are some species of *Spirogyra* in which a fusion (through connecting tubes) occurs between the contents of adjoining cells of the same filament.

Spirogyra provides merely one example of the sexual differentia-tion of individuals, which appears, as well, in some of the species of *Oedogonium* and in still other green algae. It appears regularly in an interesting group known as the "desmids." Desmids are beautiful little single-celled plants that occur in an amazing variety of sizes and shapes. In every one, however, the cell is organized into two

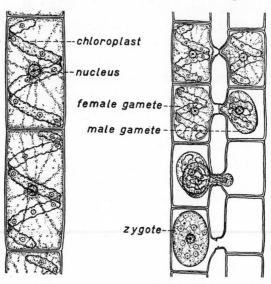

--chloroplast

-nucleus

female gamete--

male gamete -

zygote-

FIG. 12.—*Spirogyra*. On the left, an enlarged portion of the vegetative filament, showing the striking ribbon-like chloroplasts. (Some species of *Spirogyra* contain within each cell several of these chloroplasts in parallel, spiral arrangement.) On the right, conjugation of male and female filaments, with the successive stages in the process shown in sequence from top to bottom.

perfectly symmetrical halves, each half with its chloroplast and the nucleus in a position midway between the two (Fig. 13). Desmids reproduce by cell division or by a sexual fusion. Two individuals, coming to lie side by side, produce a single zygote through a fusion of the contents of the two cells.*

* The sexual reproduction of the desmids is unique among methods of repro-duction in providing a reproductive ratio of only one, for the zygote that results from the fusion of two individuals yields only two individuals of the new genera-tion. If desmids had had no other method of reproduction they could not have persisted on the earth.

The brown algae (Phaeophyceae) owe their name to the presence of additional pigments which more or less conceal the chlorophyll, producing color effects which range from olive green to a rather deep brown. With very few exceptions these forms live submerged in sea water, from which they appropriate iodine. Man takes advantage of this by utilizing the larger bodied brown algae as his commercial source of iodine.

The relationships of the group are still rather obscure. There is a general impression among botanists that they have been derived from the green algae. Along with this usually goes the notion that life originated in fresh water, and that salt-water habitats were invaded later by some lines of descent. It is quite possible, however, that both brown and green algae were derived long ago from some common ancestral stock that has since become extinct. Some of the brown algae do indeed resemble the greens in more ways than one, but the same forms include little structural peculiarities which differ rather sharply from the corresponding features of green algae.

FIG. 13.—A desmid.

No single-celled brown algae are known. At its simplest the body is a branched filament, one cell in caliber, which in size and contour is similar to the body of many green algae. The majority, however, are much larger and coarser than this, many-celled in caliber, and with a certain amount of differentiation of body tissues. Usually there is a strong, many-celled holdfast at the base, by means of which the alga is anchored in comparatively shallow water along rocky stretches of the sea coast. The rest of the body may fall into any one of a great many patterns. Stretching upward from the holdfast is a tough, leathery stalk of a length suitable to carry the upper portions of the body near to the surface of the water. These upper portions are more or less broken up into flat blades, which stretch out horizontally near the water surface and conduct most of the food manufacture. To maintain this position they may be assisted by one or more "air bladders" or "floats" (Fig. 14). Those forms which are

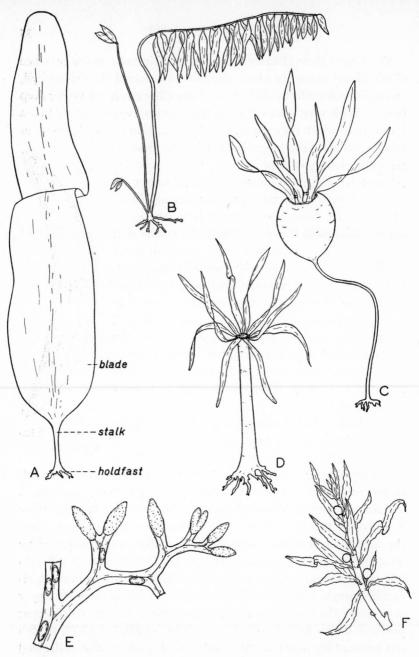

- - blade

- - - stalk

- - - holdfast

FIG. 14.—An assortment of the coarse-bodied brown algae, including: (1) "kelps," *A, Laminaria; B, Macrocystis; C, Nereocystis; D, Postelsia* ("sea palm"); (2) "rockweed," *E, Fucus;* and (3) "gulfweed," *F, Sargassum. (A–D* particularly are often many times larger than here shown.)

best equipped with long, rope-like stalks and floats can keep their blades in the surface position at both high and low tides. Those which lack this flexibility may be deeply submerged at high tide. The whole body is of an exceedingly tough consistency, adapted to resist the pounding of the waves—an adaptation that frequently fails, as testified by the great numbers of "sea-weeds" cast up on the shore after a storm.

In published accounts one can find amazing reports of the size attained by some of the larger brown algae, lengths of up to 1,500 feet being cited. Most of these reports can be traced back to the stories of the old sea captains and other travelers who have always had a weakness for impressing the stay-at-homes with descriptions of monsters of the deep. Reliable records place the maximum length at about 150 feet, but even this is a rather impressive figure, especially when compared with the body size of the other alga groups.

In contrast with the red algae, to be considered next, most of the browns "prefer" cool water. Hence we encounter the greatest abundance of them along the northern stretches of both our Atlantic and Pacific coasts, where the natives commonly speak of them as "rock-weeds" and "kelps." *Sargassum*, however, is a rather unorthodox genus. Popularly spoken of as "gulf weed," it is commonly found free-floating on the surface of the warmer waters in and near the Gulf of Mexico. Here it forms extensive tangled mats of vegetation, which are said to have been quite a menace to the old sailing vessels, as recorded in the experiences of the "Ancient Mariner" and of Christopher Columbus.

When it comes to reproduction, we find that most of the brown algae employ the ciliated swimming spore, similar in function to that of the green algae but different in structural detail. In addition, most of them imitate *Ulothrix* with a sex program which involves similar, swimming gametes. As in the green algae, however, one could arrange quite an impressive series of brown alga types to show a steady increase in size and loss of motility by the female gamete. This culminates in forms like *Fucus*, the commonest of the rock-weeds of our New England coast, in which the discrepancy in size between the two gametes surpasses anything found among green algae. *Fucus* has the additional peculiarity of discharging its enormous, non-motile eggs into the water. The *Fucus* egg, bombarded by a halo-like swarm of diminutive sperms, and set into rotation thereby, is a very impressive sight under the microscope (Fig. 15).

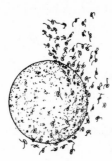

FIG. 15.—Fucus egg and sperms. (The actual size difference between the two gametes is greater than indicated in the illustration.)

The red algae (Rhodophyceae) are also a marine group, having only a few exceptional forms that live in fresh water. A red pigment completely or partially conceals the chlorophyll. The body is usually a delicate ribbon or finely-branching filament, which would be very poorly adapted to the turbulent habitat of the brown algae (Fig. 16). Instead, we find most of the reds living at quite a depth below the surface, where waters are quiet and the more serious problem is to obtain enough sunlight for food manufacture. There is also a decided preference for high temperatures. Visitors to tropical and subtropical coasts often return with attractive souvenirs, which are made by simply floating one of the feathery bodies of a red alga onto a white card, where it will dry and stick, due to the gelatinous material it contains. The "agar" of commerce, used medicinally and as an artificial culture medium for bacteria and fungi, is a gelatinous extract from the red algae.

Affinity with the brown algae is suggested by the existence of certain algae which show a combination of characteristics from both groups. Even so, botanists feel quite uncertain of the relationships.

The reproductive structures of many of the red algae reach a complexity surpassing that of any of the other groups. The group as a whole is characterized by the production of non-motile spores and sperms, which are quite at the mercy of chance water currents.*

---

* The obvious inefficiency of such methods of distribution and fertilization is hard to reconcile with the apparent success of the group. Perhaps one should conclude that indeed the red algae have failed in the past when in competition with the greens and browns in habitats that can be tolerated by the latter; that they are a success only in their own peculiar environmental niche, i.e., deep waters; and that their adaptation to this environment is the unique red pigment they possess, which may somehow enable them to get along on a more dilute sunlight than would otherwise be possible.

Sperms are floated over to stick upon and discharge their contents into tiny filamentous protrusions from the female sex organ. The

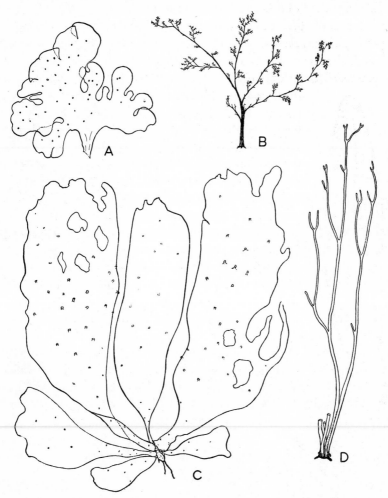

FIG. 16.—An assortment of red algae. *A, Pugetia; B, Polysiphonia; C, Gigartina; D, Nemalion.*

resulting zygote passes into a long program of cell divisions which yields a mass of spores, usually surrounded by elaborate accessory structures—but such a crude summary falls far short of the actual complexities involved.

Though most of the alga groups have left practically no fossil record, owing to the softness of their bodies, a striking exception is provided by the "diatoms." These are one-celled forms which possess a remarkable type of cell wall. Heavily impregnated with silica, it persists for an almost indefinite period after the protoplasm within has died and disappeared. Diatom populations in past ages must have reached unthinkably high numbers. Though the individual cell is microscopic in size, the accumulation of the persistent cell walls was almost wholly responsible for large deposits which today we refer to as "diatomaceous earths."

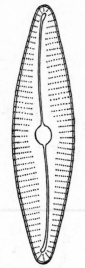

One who relies upon the naked eye misses a great part of the beauty of nature, which can be revealed only by the microscope. The siliceous external "skeleton" of many diatoms is sculptured with a series of lines, so fine and so regular that they have often been used to test the powers of definition of microscope lenses (Fig. 17). One is quite impressed upon finding that these external markings are just as true on the fossil forms of hundreds of millions of years ago as on the living forms dipped out of a neighboring lake.

FIG. 17.—A diatom.

Many diatoms grow on damp earth, but probably more in fresh and salt waters. There they are sometimes found attached to the bottom by diminutive holdfasts, occurring thus either singly or in branching, filamentous colonies. Many others float free at the surface of the water, where they form a larger part of the fresh-water and salt-water "plankton."*

---

* The plankton is a world in itself—a world populated by countless billions of organisms. Floating at or near the water surface, and bathed in sunlight, diatoms and other microscopic algae steadily manufacture food which supports not only their own bodies but also those of many dependent organisms that swarm in the neighborhood. Through microscopic herbivorous animals and a series of larger and larger carnivorous forms, the green members of this free-floating plankton support the massive bodies of sharks, octopi, and whales;

Some botanists are inclined to classify diatoms as green algae, for in their reproductive features they show some resemblance to the desmids. Others put them among the browns, on the grounds of an accessory brown pigment, which is not, however, identical with that of the brown algae.

---

while the dead bodies of all these forms, sinking to the unlighted depths of the ocean, are the only source of nourishment for a weird population of fish and other animals that never see the light.

CHAPTER IV

DEPENDENT ORGANISMS

THE spread of mankind over the face of the earth has been made possible by the repeated establishment of pioneer communities in previously uninhabited regions. These pioneers have usually been hardy souls, capable of maintaining an independent existence, of wresting their livelihood directly from nature without the assistance of any human beings outside of their own community. After these early settlers have established themselves successfully, and the local community has commenced a thrifty growth, men of a different stamp almost inevitably begin to appear. These newcomers proceed to support themselves not independently but dependently, not by exploiting nature directly but by exploiting their local fellow-men.

In most cases the newcomers are what the name implies, having migrated to the pioneer settlement from older communities. In some cases the dependent individuals are of local origin; for a few of the pioneers themselves (or their descendants) may have changed to adopt dependent habits. In either event the new, dependent individuals must have been derived from an ancestry which was at one time independent.

In the conquest of the earth by living organisms, the pioneers in any region were probably some form of green plants. For today we find that, save for a few rare types of bacteria, the only organisms that are able to live independently are the green, food-manufacturing plants. Their green pigment, chlorophyll, enables them to take energy from the sunlight and use it in manufacturing organic food from inorganic compounds that have no food value.

We have seen how the algae played the rôle of pioneers in the conquest of the various regions of the sea and of the fresh waters. Later we shall see how higher green plants, descendants of the algae, were pioneers in the conquest of the various regions of the dry land. On countless occasions the same general sequence of events must have occurred, involving the gradual establishment of a thrifty community of green plants in a previously uninhabited locality.

On all these occasions the first main act in the drama was followed with deadly certainty by the second. Given a new community of green plants, and dependent organisms were certain to appear, organisms which did *not* manufacture their own food but which, instead, by one device or another, appropriated some of the food that had been manufactured by the green plants. Time after time dependent organisms were evolved from independent ancestors and commenced to prey upon their independent relatives. And in the course of time secondarily dependent organisms were evolved—organisms which preyed not directly upon their independent relatives but upon their dependent relatives. So that today our organic world consists of a broad foundation of green plants, supporting a towering superstructure of other organisms, some of which are only remotely, but nonetheless certainly, dependent on the food-manufacturing power of the green plants.*

In the course of time three great groups of dependent organisms have succeeded in establishing themselves. Each is today widely distributed over the earth, and each is represented by tens or hundreds of thousands of distinct species.

The most conspicuous of these groups consists of the animal kingdom. This vast assemblage of a million or more species is bound together by the common property of dependency.† Animals work out their lives on the basis of an exploitation—and usually a direct exploitation—of other living organisms. The history of the animal kingdom has been characterized by a series of triumphs for those types most successful in this matter of exploitation. Man's own

---

* Theoretically there might be some cases in which independent and dependent organisms advanced simultaneously into new territory; and actually this phenomenon is illustrated by the lichens, viz., chap. vii.

† Actually a few of the simpler animals contain chlorophyll, and with it the power to manufacture a certain amount of food in addition to that which they appropriate by the characteristic animal method. Since animals almost certainly evolved from plants in the first place, it is not surprising that a few have retained to some extent this characteristic of their ancestors.

supremacy has depended upon his unusual success as an exploiter of other living things.

Another great dependent group is the bacteria. Judged by structural characteristics, these are relatives of the blue-green algae. In bacteria we find the same microscopic, one-celled bodies, with no clear differentiation of a nucleus, and simple cell division as the only method of reproduction. Again we find amazingly high resistance to extremes of conditions—in some bacteria even higher than in the best of the blue-green algae—and a corresponding ubiquity. The conspicuous difference lies in the absence of chlorophyll among the bacteria.

From these considerations one might well conclude that bacteria are dependent offshoots from the blue-green algae line; and that here we had identified the earliest of all adventures in dependency. Such may, indeed, be the truth, but there is another interpretation which is, perhaps, equally plausible. Perhaps some still more primitive group, with independence of a different type, gave rise on the one hand to the green, independent blue-green algae, and on the other hand to the non-green, dependent bacteria. Such a view is suggested by the discovery of a few simple organisms, structurally like the bacteria, but functionally capable of manufacturing food without the assistance of chlorophyll.*

Whatever their origin, the bacteria are today the real rivals of the animal kingdom. Man himself, at least in civilized communities, has firmly established his dominance over the rest of the plant and animal kingdoms; but the bacteria, while actually serving man's interests in many ways, continue to be a serious menace to his health and welfare. Further discussion of bacteria is deferred until chapter xv.

A third great group of dependent organisms is the fungi. Though this group does not dominate the organic world to the same extent

---

* Among the organisms classified as bacteria are a few (the purple and green "sulphur bacteria") that manufacture carbohydrates in sunlight by means of pigments other than chlorophyll. There are a few others (the "iron bacteria") that get the energy for food manufacture not from the sunlight, but from the changing (oxidation) of ferrous iron to ferric iron. Both of these groups can thus live quite independently of other organisms.

as do the animals and the bacteria, it is nonetheless a vast assemblage, and one of great economic significance. This consideration, together with the prominent place occupied by fungi in the plant kingdom, justify us in devoting the next few chapters to a consideration of the group.

Animals, bacteria, and fungi are the three *great* groups of dependent organisms. Most, but not all, of the known dependent organisms fall into these groups. In addition, dependency is exhibited by a few scattered members of the highest plant group, the seed plants. These dependent seed plants will be discussed briefly in a later context.

CHAPTER V

A SAPROPHYTIC FUNGUS

FUNGI are characterized by the possession of a "mycelium."
A mycelium is a plant body consisting of a system of fine,
branching filaments. Though colorless for the most part, some
mycelia contain pigments of one color or another, but never chlorophyll. In the active, working part of the body the mycelium is of
cobweb consistency, its countless fine branches spreading out over
the surface or through the mass of the material from which it is taking nourishment. In some fungi, however, a considerable portion of
the body is specialized for reproduction, and in this portion the
mycelial threads are compacted into a solid, and often a surprisingly
hard, mass, as in the common mushroom and in the bracket fungus
that is often seen growing on old trees and stumps.

In their nutritive relationships all fungi are dependent on organic
matter. Some utilize organic matter that is no longer a part of a living organism, such as bread, jelly, manure, or the dead and disintegrating body of some animal or plant. Such fungi are known as
"saprophytes." Man himself and the majority of animals hold a
similar nutritive relationship with their food sources.* Other fungi
prey directly upon the bodies of living organisms. In such a case the
fungus is referred to as a "parasite," while the organism that is being
exploited is known as the "host." Some animals (e.g., fleas, ticks,
tapeworms, hookworms) are likewise parasitic. The majority of
fungi are either strictly saprophytic or strictly parasitic, but there
are some that have the capacity for both nutritive relationships.

The general characteristics of fungi are rather simply and clearly
displayed by the ubiquitous "bread-mold" (*Rhizopus nigricans*),
found commonly on stale bread, jelly, and manure piles. Many a

---

* Since saprophyte technically means "rotten plant," the term could not
properly be applied to animals. Such animals might be said to be "saprozoic,"
though there is little call for such a term in biological parlance.

housewife, planning to salvage the remainder of a three-day-old loaf, has been distressed to find parts of it covered with fine white "whiskers." In fact, most of us have doubtless eaten bread in this condition without realizing that we were getting more than we had bargained for. Since the microscopic spores are being produced copiously and released into the air by the almost innumerable patches of mold that are growing somewhere in the neighborhood, and since the spores are so light as to be readily carried by the slightest of air currents, it follows that a goodly number of them are circulating in the atmosphere of every household. Repeatedly they are lighting upon objects of all sorts. It is only, however, when they chance to light upon a favorable nutritive medium (under the proper conditions of humidity and temperature) that they "germinate"* to produce the mycelium body of the fungus.

During the first few minutes of its exposure on the kitchen table the loaf of fresh bread is likely to receive one or more of these *Rhizopus* spores. Lodged on this relatively warm and moist nutritive medium, the spore will, within a few hours, germinate to produce the first microscopic filament of the new mycelium. Growth is rapid, but the mycelium is so minute and colorless that it is usually only after two or three days that it has reached proportions discernible by the naked eye.

The main mycelial body grows only along the surface of the bread, branching repeatedly and spreading in an ever widening disk from the point of its origin. Here and there, for nutritive purposes, it sends down into the substance of the bread little root-like processes that are of smaller caliber and more copiously branched than the main mycelium. Through the surfaces of these little branches, water and organic material are absorbed from the bread and passed upward to provide for growth of the main mycelium. Thus the purely vegetative life of the fungus is simple enough, since there is no problem of food manufacture or of the building of a complex body and no

---

* Strictly speaking, "germination" refers to the early stages in the development of a new vegetative body from the single-celled spore or the single-celled zygote. A somewhat looser usage of the term applies it, as we shall see later, to the events connected with the emergence of the young plantlet from a seed.

particular difficulty involved in the transport of materials within the body.

Reproductively, however, the bread mold is more complex, illustrating all of the three major types of reproduction that occur in the plant kingdom. Simplest of the three is reproduction by "vegetative multiplication," in which the unmodified vegetative body itself effects a multiplication of individuals. If one were to detach a fragment of the mycelium and transplant it onto a new nutritive substratum, this fragment would constitute a separate individual and would soon develop to whatever size was made possible by the amount of food available in that locality. Even when left to itself the spreading fungus mycelium often becomes a series of detached individuals through the starvation and death of intervening parts.

If such vegetative multiplication were the only method of reproduction, however, this fungus would be rare rather than ubiquitous. Fragments of mycelium are rarely effective as agents of distribution, for any very lengthy trip through the air would dry and kill them. Such a conspicuously successful species must possess a more efficient means of distribution.

Usually, when the fungus is only a few days old, numerous vertical branches will arise from scattered points on the mycelium. Each of these branches is at first a very simple affair, uniform in caliber throughout its entire length. Very shortly, however, a spherical swelling appears at the tip of each branch. This sphere continues to enlarge until it is eight or more times the diameter of the branch that carries it. Up to this point the sphere contains merely the pigmentless cytoplasm and nuclei that are characteristic of the mycelium itself. At such a stage the growth of mold would appear under the hand-lens as a miniature thicket of delicate stems tipped by snowy white balloons.

Within another day, the spheres turn black. This is due to the fact that the contents of each sphere are cut up into hundreds of tiny spores, each equipped with nucleus and cytoplasm and surrounded with a dark brown wall (Fig. 18). Now our loaf of bread advertises its moldiness to the naked eye, for the black color of the ripe spore-cases, tiny as they are, makes them stand out clearly against the white background of the bread. It is due to this fact that the popular name of "black mold" is applied to this fungus almost as frequently as is "bread mold."

In these reproductive organs, as in those of all higher plants and animals, we can distinguish "fertile" and "sterile" elements. In the case in hand, the spores are the fertile elements, for they are the cells which actually develop into the individuals of the next generation. Of the sterile and purely accessory elements there are two, the spore-case (i.e., the wall surrounding the entire group of spores) and the vertical stalk which bears it. Though these latter are sterile, they nonetheless contribute to the efficiency of the program of spore re-

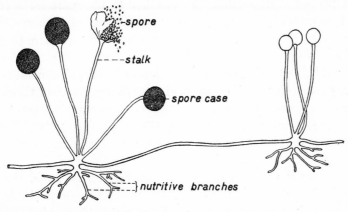

Fig. 18.—A magnified diagram of the bread-mold mycelium, together with the structures involved in spore reproduction.

production. The spore-case surrounds and protects the spores in their tender developing stages. The stalk elevates the spore-case and thereby increases the likelihood that the spores, when released, will be caught up by chance air currents and carried to some distance. Throughout the fungi (most of which live and release their spores in a medium of air) these two sterile elements recur in a variety of forms, sometimes quite elaborate.

When the spores are ripe, an upward pressure of the watery protoplasm within the stalk comes to bear upon the contents of the spore-case. This pressure, transmitted through the mass of spores, stretches the outer wall (i.e., the spore-case proper) until at last it reaches the limit of its elasticity. At this point it bursts, forcibly ejecting most of the spores, and thus catering further to their effective distribution. Similar devices appear among higher plants, providing for an active discharge of spores or other reproductive units which may act as distributing agents for the species.

Unlike the spores of the water-living algae, those of the bread mold are light and provided with a wall which resists desiccation. In other words, they are adapted to distribution in a medium of air. The tiny bit of protoplasm within the spore wall is in a state of dormancy. Animation is not completely suspended, but life goes on at an extremely slow rate, so slowly that even the meager supply of nourishment present will maintain the viability of the spore for several years. Whether it is a day or a year after the time of discharge, that spore which lights on a suitable substratum will germinate to produce a new mycelium.

The third method of reproduction is sexual. In their general vegetative characteristics all bread mold mycelia are alike, but the manner of sexual reproduction reveals that actually there are two types of mycelia, commonly referred to as the + (plus) strain and the − (minus) strain. Neither strain by itself will ever initiate sexual reproduction, for here there is clearly a sexual differentiation of individuals, and the two strains must co-operate in the production of zygotes. When a strand of mycelium from the plus strain grows close to one from the minus strain, the two put out tiny side branches which come into contact at the tips. At a short distance behind the tip of each branch a wall is laid down. The two small masses of protoplasm which are thus cut off at the tip of each branch are to act as gametes. In this case, however, there is nothing in the relative size or activity of the gametes to betray which is male and which is female. Both stay where they are, but the contiguous end-walls of the two dissolve away and allow the two little masses of protoplasm to fuse into one. This fusion product is, of course, the zygote. Soon it enlarges and lays down about itself a rather irregular, heavy black wall (Fig. 19).

Ordinarily this zygote remains dormant for months, later to be awakened if moisture and warmth are provided. Under such conditions the zygote wall is softened, and the contained protoplasm pushes through. At once it produces a vertical stalk and spore-case.

Each of the many spores produced from this spore-case has the power to produce a new mycelium.*

The three possible life-cycles shown by this fungus represent the three fundamental types of reproduction for living organisms in general.

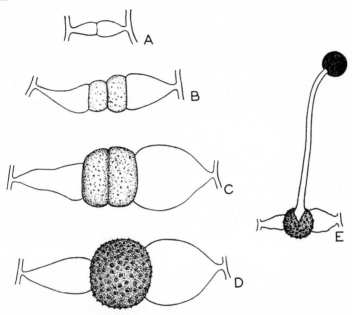

FIG. 19.—Sexual reproduction in the bread mold. *A–D*, successive stages in zygote-formation; *E* (taken from a related fungus), germination of zygote to produce stalk and spore-case.

In the first and simplest type, which, in the plant kingdom, is usually called "vegetative multiplication," there is no setting-aside of specialized reproductive cells. Instead, a part of the ordinary vegetative body is transformed directly into the individual of the next generation. Vegetative multiplication is at its simplest in the unicellular organism, where ordinary cell division is, at the same time, an act of reproduction. In the multicellular organism it is

---

* Since the behavior of the zygote of *Rhizopus* itself is decidedly atypical, the account in the paragraph above is borrowed from the more typical life-cycle of *Mucor*, a closely related form.

usually not one alone but a group of vegetative cells which becomes the new individual. In this case, too, the new individual owes its origin to a series of cell divisions. The difference between such a case and that of the unicellular lies more in the matter of separation from the parent-body. In the unicellular, cell division itself functionally separates the body of the new individual from that of the parent. In the many-celled organism, a group of new cells which may result from a series of cell divisions remains functionally a part of the parent-individual and continues to co-operate with the rest of the body until such time as it becomes separated through the action of some other agency. Through accidental severance or the death of intervening parts, the group of vegetative cells may become isolated, functionally separated from the parent-body. If, then, the isolated group of cells has not already become too highly and irrevocably specialized as a subordinate part of the parent-body it will itself continue to grow as a separate individual. Hence many of the lower plants, whose body parts are not highly specialized, have a large capacity for vegetative multiplication. As we pass to the higher and more specialized plants, we find fewer cases of vegetative multiplication; but even among the highest of plants we often find that certain of the body parts, apparently less specialized and subordinate than the rest, are capable of effecting vegetative multiplication if separated from the parent-body and placed in appropriate conditions. Specialization and subordination of body parts has, of course, been carried much farther in the animal kingdom, so that among the highest animals reproduction by vegetative multiplication is a practical impossibility.

In the second of the fundamental types of reproduction, spore reproduction, we encounter a characteristic combination of several features: (1) Though the body of the individual may be many-celled, the spore is a single cell, capable by its own independent action of producing a new individual.* (2) Some device is present for the regular

---

* Among some fungi there are exceptional types of spores which are two-celled or even many-celled. In these cases, however, each of the cells which goes into the make-up of the spore is capable by itself of producing a new individual. Here, apparently, we are dealing with a compound spore.

release of ripe spores into the surrounding medium of air or water. (3) Spores are more or less specialized as agents of distribution, being equipped in such manner as to encourage their wide dispersal through the medium. (4) In most cases a single parent-individual produces its spores in large numbers.

Sex, the third of the fundamental types of reproduction, is readily characterized by the feature of fusion. The details of sexual repro-duction vary tremendously in the plant and animal kingdoms, but these details are merely accessories to the fusion of two cells (and, most significantly, of the two nuclei within those cells), commonly from different sources.

CHAPTER VI

PARASITIC FUNGI

THOUGH the bread mold may prove quite annoying to the housewife, though it may also cause considerable losses in stores of such vegetables and fruits as sweet potatoes, apples, peaches, and strawberries, and though various saprophytic fungi may, at times, impair or destroy certain items of man's property that are composed of organic material, the fungi of more serious economic importance are the parasites. In the notorious "ring worm" or "athlete's foot," and in a few rare diseases of the throat, we have examples of parasitic fungi that are making a direct attack on man's body. Similarly there are a few that attack the bodies of other animals. But the vast majority of the victims of parasitic fungi are the green plants, and when the green plant happens to be an important food source for man the fungus may be of tremendous economic importance.

A classic example of the parasitic fungus is the one producing the disease known as "late blight of potato."* We shall start the story of its activities with the time when one of the light, air-borne spores of this fungus is blown to lodge on the leaf of a young potato plant in the spring.

Now there are some parasitic fungi that confine their attack to the skin regions of their hosts (i.e., the plants and animals which are their victims). For such external parasites there is no serious problem of penetration, since mere lodgment on the surface of the host puts them in direct contact with the tissues which are to provide their food. The skin diseases which such parasites produce may be moderately destructive, but are seldom fatal. It is the internal parasite which is usually more to be feared.

In the case of the internal parasite, mere lodgment of the spore

---

* The fungus itself carries the outrageous scientific title of *Phytophthora infestans*.

on the surface of the host is no guarantee that infection will result. For true infection, penetration must be effected, and this is usually no simple matter, for nature has endowed most plants and animals with skin tissues which are highly efficient in preventing the entrance of internal parasites. Penetration may occur through one of three avenues, depending on the nature of the fungus and the nature of the host. (1) Some fungi effect entry only through wounds. Knowing this, the tree surgeon, after pruning off a branch, hastens to coat the exposed surface with some substance which is physically or chemically an effective obstruction to the entrance of wood-destroying fungi. (2) Other fungi make use of the natural openings in the host, the tiny breathing-pores which are scattered over most of its surfaces. In such cases it is not necessary that the original lodgment by the wind be in, or even very near to, the breathing-pore, for the young mycelium which emerges from the spore can grow along the surface for some distance before reaching a breathing-pore. (3) Unfortunately for the prospective hosts, however, there are still other fungi capable of producing strong digestive fluids with which they eat their way directly through intact surfaces.

Though probably the commonest thing is for a fungus to be endowed with only one of these three devices for penetration, there are some fungi that can evidently rely upon more than one, and the reports are that the fungus producing late blight of potato can penetrate by any one of the three methods. Once within the tissues of the host, our fungus, having by this time pretty well exhausted the food that was stored within the spore, must immediately tap a new food supply. The main mycelium of the parasite usually grows not in, but between, the cells of the

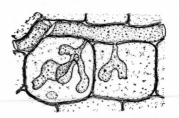

FIG. 20.—Portion of an internal parasitic fungus, showing main mycelium wedging between the cells of the host, and nutritive branches actually penetrating the host cells.

host. In leaf tissue this presents no problem, for there is already established in the leaf a connected network of air spaces between the cells. But the main mycelium is not all. Just as the main mycelium of the bread mold pushes nutritive branches into the substratum, so also does that of the parasite branch out with tiny processes which actually enter the host cells to digest and absorb much of their living contents (Fig. 20).

Parasitic fungi are surprisingly specific in their tastes. Very commonly a given fungus will confine its attack not merely to a single host species but often to only one or a few varieties of that species. Furthermore, its attack is usually restricted to a particular type of tissue within the host. Other tissues may be traversed by the main mycelium, but are not exploited as a source of food. The fungus under discussion may grow through all parts of the potato plant, but for the most part it attacks and destroys only those rather soft cells of the leaves that contain the chloroplasts. Pushing on rapidly from one leaf to another, it may thus wreck the food factories of the entire plant within a week or so after the time the original infection occurred.

FIG. 21. — Spore-bearing branch of the fungus responsible for "late blight," emerging through a breathing-pore on the lower surface of a leaf of its host.

Just as the bread mold, after a time, changes the local substratum enough to create the stimulus that makes for spore production, so it is also with all parasitic fungi. Again spores are produced on stalks, but in this case the stalks (or most of them) grow downward from the mycelium. Though the leaf may be pretty well "gutted" by this time, its general framework is still intact, including the skin layer. The stalks which are to produce spores, apparently unable to penetrate the skin itself, emerge through the tiny breathing-pores, and most of these are on the lower surface of the leaf. Spore-cases are seldom produced. Instead, the stalk branches repeatedly, and at the tip of each branch a single spore is cut off (Fig. 21). Within a day or so a single infected potato plant may thus fill the air with many thousands of spores, so that other plants in the neighborhood are almost certain to be infected. Often the successful infection of one young plant in spring may leave the entire potato patch devastated before the end of the growing season.*

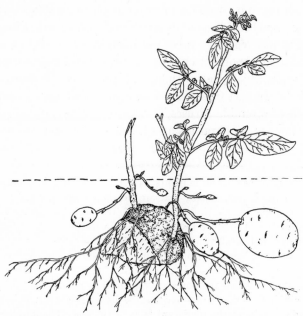

FIG. 22.—Diagram to show how potato plants arise (by "vegetative multiplication") from a bit of tuber, and how these plants, in their turn, produce a new generation of tubers on underground branches from the main stem.

This fungus is a serious enemy of man whether or not it completely kills the potato plant which is its host. The potato of commerce is, technically, a "tuber." It is a specialized underground branch from the main stem in which a great deal of food is stored in the form of starch (Fig. 22). The healthy potato, during the growing season,

---

* In addition, though apparently rarely, the late-blight fungus engages in sexual reproduction, in which unlike male and female elements unite to form zygotes. This is apparently of little or no value to the species, however, since distribution is accounted for by the spores, and wintering-over from one season to the next is usually accomplished by bits of mycelium which stay alive within the tubers from infected plants.

manufactures much more than enough food to supply current needs, and most of this excess is stored in the underground tubers. The blighted plant, even though it may escape complete destruction, has lost so much of its food-manufacturing tissue that there is little or no excess food to be stored, and the crop which man digs up consists of useless "marbles" instead of fully developed tubers. (In addition the fungus may reach the tuber itself and induce a rot.)

A large and important chapter in the story of man's conquest of nature is that which deals with his warfare against the diseases of his crop plants. In part these diseases are due to bacteria, in part to the mysterious "filterable viruses,"* occasionally to animal parasites, and at times to no parasites at all, being apparently nothing more than "functional diseases."† The majority of serious plant dis-

---

* There are some diseases in plants, as well as some very important diseases in man and the lower animals, that are clearly infectious, but for which the biologist has been unable to find any visible parasite. An extract from a diseased organism will produce the same disease when it is introduced into the body of a previously healthy organism. Yet no visible evidence of life appears in this extract even when it is examined under the best of microscopes. Furthermore, this extract retains its infective powers even after it is passed through a porcelain filter in which the pores are too small to permit the passage of the smallest visible bacteria. To assume that the disease is due to a lifeless chemical is to overlook some of the significant facts of the case. The infection can apparently spread through an infinite series of hosts. In such a program any lifeless chemical would soon be so diluted as to lose its toxicity (capacity for poisoning). Whatever the real nature of the infective agent, clearly it has the capacity for self-multiplication (i.e., is "autocatalytic"). Some are inclined to speculate that we are dealing in these cases with miniature bacteria, but pending a clearer revelation of the nature of these substances, biology refers to them merely as the "filterable viruses."

† Through most of human history, man attributed disease to supernatural agencies. His whole mental attitude was strongly flavored by a sense of insecurity which was inevitable in a world oppressed by invisible, mysterious forces that often struck without warning at his life or his health. Very naturally he sought such reassurance as was available in religious and semi-religious ceremonies which sought to propitiate the threatening agencies. The result was only too often either a failure to gain the desired reassurance or a gaining of it at the cost of a decided clouding of his picture of the natural world.

The "germ theory of disease," which was effectively publicized by Pasteur and others during the closing decades of the nineteenth century, had a tremen-

eases, however, are attributable to parasitic fungi. In man's fight against plant disease, there is usually no such effort to save the individual patient as in the case of diseases of man himself or his more valuable domesticated animals. Man stands quite ready to sacrifice an individual plant or so, if only an epidemic can be prevented and the bulk of his crop saved. The two commonest tactics which man uses in this warfare are the application of "fungicides" and the breeding of resistant varieties.

As the name implies, a fungicide is a fungus killer. Experimentation may demonstrate that a certain chemical or combination of chemicals will kill the spores or young mycelium of the fungus without at the same time working any serious injury on the superficial tissues of the host plant. A solution of the chemical, mixed with

---

dous influence in the emancipation of the human mind. Man could now hope to understand the true cause of disease and to combat it.

It is unfortunate that a successful new idea is almost always pushed farther than the facts of the case warrant. The impetus that was originally given by Pasteur has resulted in the popular impression that *all* diseases are the result of "germs" or "microbes." The modern biologist realizes the fallacy of this impression.

The "healthy" organism is one that is so adjusted to its environment as to keep its body alive with a minimum expenditure of energy, as well as to perpetuate, multiply, and distribute the species. All other organisms are to be regarded as "diseased" or in a "pathic state."

Whatever an organism (or any part of an organism) is, depends upon an interaction of the factors of heredity and environment. The factors of heredity are provided in the nucleus of the zygote from which the organism starts, and are reproduced in the nuclei of the body cells, all of which have been derived from the zygote by the remarkably regular process of mitosis. The factors of the environment are living ("biotic") as well as non-living (physical and chemical). The healthy state depends upon a favorable combination of all these factors. Disease results from the unfavorable (or "pathogenic") action of any one of them.

If a corn plant is derived from a zygote that chances to lack only one of the many hereditary factors that are necessary for the production of chlorophyll, that plant is diseased. It starts life as a white seedling, but, since it cannot manufacture food, dies as soon as the food that was stored in the seed is exhausted. If a human individual lacks one of the hereditary factors necessary for normal brain development, that individual is diseased, for he has an equip-

some adhesive substance, is then sprayed upon the surfaces of the crop plants. Due to rain, and for other reasons, the coating of fungicide gradually disappears, so that several sprayings are often necessary during the growing season. Of course this procedure confers no benefit upon the plant that is already infected, but usually prevents new infections. The spore of the fungus, lodged on the surface of its host, puts out the young mycelium only in the presence of a little moisture. On a sprayed surface, this moisture will have dissolved a little of the fungicide, and the result will be a poisoning of the young mycelium. In the past the control method most generally employed for late blight of potato has been spraying with a fungicide known as Bordeaux Mixture.*

For some crops, such as wheat, the plantings are so thick and so extensive that a program of spraying has proved impractical.† Control of diseases of such crops has been sought through the breed-

---

ment which meets the demands of life very inefficiently. Without the assistance of his fellow-men he would probably fail to stay alive.

If a man falls down and breaks his leg he is diseased, until such time as the member is adequately repaired. If a man gets drunk he is diseased, until such time as his system throws off the alcohol and readjusts itself. Inadequate nutrition produces disease in all dependent organisms; for, to maintain itself in a healthy state, the dependent organism must secure a fairly continuous supply of energy and material that were previously incorporated in the bodies of independent organisms. Beyond this there is always the possibility of some maladjustment appearing—as the result of unknown causes—in the structure or function of such a complex machine as the living organism. It is diseases of this last category that have often been referred to as "functional diseases"—a rather unsatisfactory title, inasmuch as all diseases must be thought of as functional disturbances in the broadest sense.

If a man is partially devoured by a lion he is diseased. If his friend gives him a black eye he is temporarily diseased.

It is only *one* of the several great categories of disease that can be attributed to small organisms that act adversely on the body of the host. For the most part, diseases of this type are the "infectious diseases," for, the disease arises in a new host following transfer of one or more of the disease-producing organisms —"pathogenic organisms," as they are often called—from the original diseased host.

* This consists of a mixture of copper sulphate, lime, and water.

† Spraying from aeroplanes has been tried, but to date has proved unsatisfactory.

ing of disease-resistant varieties. Proceeding on a knowledge of the laws of heredity, breeders have in some cases been able to separate out from a mixed population a variety which happens to possess a combination of hereditary characteristics that makes it immune to the disease in question. In other cases they have succeeded, through crossing (hybridization), in combining into one variety the disease-resistance of a commercially undesirable variety with the desired hereditary qualities of a susceptible variety. Both procedures involve breeding programs that must be arranged and guarded with extreme care through several generations before the breeder can put on the market a new variety which, while possessing desirable commercial features, at the same time is able to grow without succumbing in a region in which the parasitic fungus is prevalent. Already man has developed varieties of cereals that completely or almost completely resist the attacks of some parasitic fungi. It should not be thought, however, that the entire problem has been solved or is ever likely to be solved once and for all. Many parasitic fungi are already prevalent in this country, and new ones are frequently added to the list, some arriving as unwelcome immigrants from other parts of the world, and others apparently being evolved anew from fungus ancestors of a different type. From this it can be seen that, though progress has indeed been made in the past, there remain plenty of practical problems to be solved by the plant breeders of the future.

The development of disease-resistant varieties has not been limited to cereal crops. In many other crops, such as our potato, spraying is arduous, not altogether certain, and needs to be repeated annually. The development of disease-resistant varieties is a far more satisfactory method of control, for the results are relatively permanent and relatively fool-proof. So we find repeated instances in which the temporizing methods of spraying have been succeeded by a more lasting solution of the problem through the production of resistant varieties.*

---

* Though spraying and the production of disease-resistant varieties are the control measures that are most widely employed today, there are others which play quite significant rôles.

Some parasitic fungi are carried on seeds from one host generation to another.

There is another parasitic fungus that merits our attention, not only because it is tremendously important in the economic sense but also because it illustrates a remarkable biological phenomenon which recurs among many other parasitic plants and quite a few parasitic animals. The fungus is the one responsible for the disease known as "wheat rust,"* and the phenomenon is that of "alternate hosts." The complete life-history of the wheat rust fungus, as it is understood today, includes quite an array of fantastic features. It will serve our present purpose to give a decidedly simplified account.

Man circumvents this in part by refusing to use for planting any seeds that have come from an infected crop. Often, however, there is a suspicion of contamination attached to all seeds that are available to him. Where the fungus is transmitted by spores lodged superficially on the seed coats, man can often defeat the fungus by a brief dipping of the seeds in some chemical. Where the fungus is transmitted by mycelium within the seed, a more difficult method of seed treatment is called for. Even here, however, man is beginning, in some cases, to learn the trick of heating seeds to a degree and for a length of time that proves fatal to the fungus without killing the seed itself.

Some fungi winter over as zygotes within the dead tissues of the last year's crop. Here methods of sanitation are called for, e.g., removal and burning of all refuse left by the previous crop.

Particularly elusive are those parasites that winter over in the soil itself. In a limited number of cases chemical treatment of the soil has proved of some value. In most cases, however, soil treatment is exceedingly dangerous, for it destroys or seriously alters the population of soil organisms (i.e., the useful bacteria, fungi, and some animals) which has been steadily at work putting the soil into the condition that is favorable for the crop plant. Such problems can be solved to an extent by "crop rotation," in which soil known to harbor a particular parasite of a particular crop is either allowed to lie fallow or is planted with some other crop for as many years as are needed to cause the parasite to die out in the soil.

Quarantine is an obvious method of attempting to prevent introduction into a region of a new and dangerous parasite that has been working its havoc elsewhere. The government spends a great deal of money in establishing quarantine barriers around (and even at points within) the country. The thoughtless motorist is often more annoyed than favorably impressed by this service, which may delay his car briefly for the purpose of inspecting its contents.

Another method of disease control, of limited application, is brought out in connection with the disease known as "wheat rust," which is considered next.

* The scientific name of the fungus itself is *Puccinia graminis tritici*.

During early summer, in a wheat field of many thousand acres, a single plant may be harboring a parasitic fungus which is pushing its main mycelium between the cells of its host and consuming a good many of them through the action of its small absorptive branches. The time arrives when parts of the mycelial body commence a development which is to culminate in the production of spores. A clumping of mycelial threads occurs which forms a tiny mat of fungus tissue just below the skin layer of the wheat plant. From this there arise thousands of microscopic stalks, lying compactly side by side, and together pushing up the skin of the wheat into a blister of pinhead proportions. Each stalk cuts off at its end a single, one-celled spore. Continued upward pressure of the fungus mass at last bursts the blister and exposes this tiny patch of spores. The individual spore, seen under the microscope, has a relatively faint pigmentation, but the mass of spores which breaks through the surface of the wheat

plant appears to the naked eye as a distinctly rust-colored spot, hence the name of the disease. In the course of a few days, thousands of these tiny rusty patches may crop out on our infected wheat plant (Fig. 23).

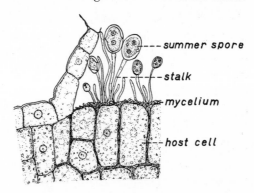

We will refer to these spores as "summer spores," naming them for the season of the year in which they are produced and in which they act.

Fig. 23.—Section through the edge of a "blister" produced by eruption through the skin layer of the wheat plant of a mass of "summer spores," the distributing agents of the wheat rust fungus.

Their action is just what we might expect. Loosened from their stalks and carried by the wind, some of them reach neighboring wheat plants. There they produce a new generation of mycelia, which enter the breathing-pores of the new hosts and establish themselves as internal parasites, similar in every respect to the parent mycelium.

This cycle may repeat itself frequently. The growing season of the wheat provides time for quite a series of crops of these summer spores, and each crop has ready accessibility to new hosts. Our single infected wheat plant of the early season may thus prove the source of an epidemic which spreads to millions of wheat plants before the time of harvest. The wheat crop may be impaired in quality or quantity, or even destroyed completely so far as its commercial value goes.

Two individual organisms with identical hereditary qualities may differ greatly in their expressed qualities if they have developed in significantly different environments. The same individual may change radically if exposed to a change of environmental conditions, particularly if this environmental change occurs during or before some period of active growth in the life of the organism. For most organisms the environment is the light and the air and the soil and all those things that we usually think of as making up the world about us. For an internal parasite the environment is provided by the tissues of its host. It follows that any significant change in the character of the host is likely to be followed by a change in the character of its internal parasite.

Toward the end of the growing season there is a waning of those conditions which have been favoring active food manufacture and growth in the wheat plant. The tissues of the wheat plant are now old, their life-activities slowing down. The world in which the fungus has been living starts to change its character, and in response the fungus, too, behaves differently. The same mycelium that produced summer spores before now produces what we will call "winter spores," naming them not for the season in which they are produced (late summer or early fall) but for the season in which they exercise a highly significant function.

Just as for the summer spores, the winter spores are produced singly on little stalks and emerge in small patches through the skin of the wheat, appearing on the stubble (stems) as well as on such leaves as may remain. These patches are black to the naked eye; the individual winter spore shows a heavy, brown wall under the microscope. Unique among the spores that we have encountered, this spore is two-celled, but, as we shall see, the two cells are destined to act quite independently of each other in the production of new mycelia (Fig. 24).

The extremely heavy wall on the winter spore betrays to us the fact that this is the stage in which the fungus winters over. Attached to the stubble the winter spore protects its two dormant cells until the coming of spring. In early spring each of the two cells puts forth a mycelium, which takes no nourishment, either parasitically or saprophytically, from its environment but subsists merely upon such food as has been stored within the spore itself. Under such a limitation it follows that this mycelium, at the stage of full development, is a dwarf by comparison with most other mycelia. At its free end it becomes divided into four cells, and each

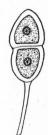

FIG. 24.—The two-celled winter spore of the wheat rust fungus. (The dark brown pigmentation of the heavy wall is not shown in this figure.)

of the four puts forth a single stalk, bearing a single tiny spore (Fig. 25). These spores we will call the "early spring spores." Their action proved quite amazing to the botanists who first discovered it, and it continues to be amazing to modern botanists even though they know of many other similar cases.

As usual, chance alone determines where the early spring spore will be blown, but of all the types of plants on which it might be lodged there is only one that it can exploit, and that is *not* the wheat plant. Only when the spore reaches a barberry bush will it

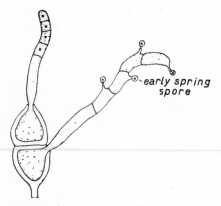

FIG. 25.—Winter spore of the wheat rust fungus, producing the tiny mycelia, which in turn produce the early spring spores.

early spring spore

establish an internal parasitic mycelium, for the barberry is the alternate host on which this fungus is just as dependent during one part of the year as it is dependent on the wheat during another.

In the course of time this mycelium, too, produces its characteristic spores, "spring spores" in this case. In spore production the mycelium within the barberry leaf develops a mass of stalks, growing compactly together, and pushing through the skin on the lower surface of the leaf. The general program resembles that which was followed in the production of summer spores and winter spores, but the final result is more picturesque. Instead of a single spore from each stalk, there is a linear series of spores, cut off successively and clinging together in regular formation. The entire "blister" is usually a perfect circle in this case, commonly known as a "cluster cup," of which many usually appear on the same leaf of the infected plant (Figs. 26, 27).

FIG. 26.—Slightly enlarged sketch of an infected barberry leaf, showing four groups of "cluster cups."

Once again spores are released, once again chance determines their distribution, and once again there is only the one host that they can exploit. Such spring spores as chance to light on young wheat plants will produce an internally parasitic mycelium like that with which we started. The cycle is at last complete. In running its course it has perpetuated the protoplasm of this fungus through three distinct types of mycelia and four types of spores, and, in truth, the botanist could point out still other complexities which we have overlooked. How this particular fungus, as well as many of its close relatives, ever evolved this strange method of exploiting alternate hosts is a question on which there has been much speculation and little elucidation.

Though the most promising method of controlling wheat rust is the breeding of disease-resistant varieties of wheat, there is another line of attack on which the United States government has already spent millions of dollars, with moderately successful results. Taking a cue from the dependence of the fungus upon the alternate hosts, the government launched a campaign for the eradication of barberry* from wheat-growing districts. In some regions the eradication has been fairly thorough and the beneficent results already ob-

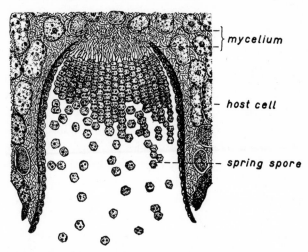

FIG. 27.—Enlarged longitudinal section through a single "cluster cup"

served. In others, the campaign has proved less successful, owing (among other things) to the large amount of barberry that has "escaped from cultivation" and is growing wild in wooded areas.

---

* The so-called "American barberry" is the culprit, the "Japanese barberry" being quite inoffensive.

CHAPTER VII

THE SUBDIVISIONS OF FUNGI

THE few examples that we have considered may have served to illustrate the general features of fungi, but they leave us with no adequate impression of the tremendous range of variability that characterizes the group. Botanists have already distinguished fully 50,000 species of fungi, and almost certainly there are many others, busily exploiting the living hosts and dead organic material which exist in remote quarters of the globe. Within this array, though the nutritive relations and the form of the vegetative body itself show no more deviation from the conventional pattern than might be expected, the course of the life-cycle and the form of the reproductive structures displays the weirdest and most unpredictable diversity. One gets the impression that the higher plant groups, in achieving their complexity, have inevitably sacrificed the plasticity or capacity for variation which existed among their simpler ancestors; while algae and fungi—many of them, at least—have retained the comparative simplicity and plasticity. One gets the further impression that the world presents a far greater variety of habitats for fungi than for algae, since differences in hosts and organic substrata can be much greater than the differences in substrata and media which nature presents to the algae. If, then, the algae and fungi were assumed to have, in some sense, an equal capacity for variation, one could conclude that the less variable world of the algae evoked less of this capacity than did the more variable world of the fungi. Whatever the speculation, the fact itself is apparent that fungi display an amazing variation, at least in the matter of their reproductive devices.

Three great subdivisions of fungi are recognized, the "Phycomycetes," "Ascomycetes," and "Basidiomycetes." Of these the first, which is much the smallest in number of recognized species, stands apart rather sharply from the other two, which are bound together by more obvious resemblances.

Phycomycetes means "alga-like fungi."* A comparative study of the filamentous bodies, the spore production, and particularly the sex structures and behavior, leaves little doubt that Phycomycetes are descendants from green algae.† Most Phycomycetes have what we call "coenocytic" bodies, meaning that there are no cross walls in the mycelium, which contains instead a continuous strand of cytoplasm, spotted with nuclei. This coenocytic condition likewise occurs in that particular subdivision of green algae which the Phycomycetes most closely resemble.

Of the three fungi which we have discussed, two (the bread mold and the late blight fungus) were Phycomycetes. This choice was in the interests of simplicity only. Actually the Phycomycetes, as might be inferred from the small size of the group, contain fewer forms of economic importance than do the other two groups.

There are indications that the actual origin of Phycomycetes from green alga ancestors took place under water. What appear to be the most primitive of the Phycomycetes are water-living forms, attacking many algae as well as fish and other aquatic animals, and distributing themselves by such ciliated spores as characterized their alga ancestors.

The aerial Phycomycetes include a series of saprophytic molds, such as the bread mold itself, a series of internal parasites, like the late blight organism, which attack grapes and many vegetables, and an interesting group which exploits insects. Many of us have noticed dead bodies of house flies adhering to window panes and mirrors. If we were observant our eyes caught a mysterious whitish ring or "halo" on the glass around the insect. The halo is actually composed of spores which have been discharged forcibly by the fungus,

---

* The two components of the term are easily recognizable, the "phyco-" being from the same root as the "-phyceae" which terminated the titles of all the alga groups, and the "-mycete" having an obvious reference to the characteristic mycelium of the fungus.

† The origin of Ascomycetes and Basidiomycetes is much more perplexing. The hypothesis that they, in turn, were derived from the Phycomycetes is far from satisfactory.

which had previously consumed many of the internals of the fly without affecting its external appearance.

Ascomycetes means "sac-fungi," the group being characterized by the possession of the so-called "ascus" or sac. The ascus, at first a simple cell, undergoes a characteristic series of nuclear and cytoplasmic divisions which yield at last a group of (usually) eight spores, the spores being retained within the old sac-like cell wall (Fig. 28). At one stage of botanical knowledge, though this development of the ascus itself was pretty well understood, antecedent events remained obscure. Now it is known that the single nucleus which is the progenitor of the eight spore nuclei has been derived (directly or indirectly) from a sex act, i.e., from a fusion of "male" and "female" nuclei. The "sex organs" which are responsible for this are often so similar to ordinary wefts of the vegetative mycelium, and are usually so well buried within the tissues of the host and of the fungus itself, that only an expert can identify them. Even today there remain quite a number of Ascomycetes in which certain identification of the sex apparatus has not been made. In addition to the spores within the ascus, most Ascomycetes also produce other spores by simpler methods at some stage of the life-cycle.

FIG. 28.—A typical ascus, the sac-like spore container that characterizes the Ascomycete group.

In terms of the number of species, Ascomycetes are the largest of the three fungus groups. One family of Ascomycetes, the so-called "powdery mildews," is responsible for an amazingly large number of "skin diseases" in plants. Here the fungus is strictly an external parasite, the mycelium spreading over the surface of the host and exploiting for nutrition merely the skin layer of the latter. A very common example is the "lilac mildew," in which the mycelium produces what are apparently dusty patches on the leaves—patches which the uninitiated owner of the lilac bush often attempts vainly to rinse off with water (Fig. 29). Other Ascomycetes are responsible for more serious skin diseases, as in the "scabs" and resultant splitting of the skin which we so often see in apples and pears.

The group also furnishes many internal parasites; forms which cause the rotting of fruits and vegetables, either before or during their transit to the market; forms which destroy whole plants by plugging up their vessels for water conduction, or by actually rotting away their entire root system; and a form that is responsible for the notorious "chestnut blight"—a disease which has mounted to

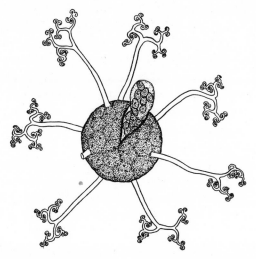

Fig. 29.—Lilac mildew. The "dusty patches" seen on the surface of an infected lilac leaf are composed of the mycelium of the fungus together with spores which are produced in chains at the end of short stalks. Here and there on the dusty patches one may notice black bodies, barely large enough to be identified with the naked eye. These black bodies are ascus-containers. One of them is shown above, with its remarkable "appendages" (function uncertain) and the ovoid asci starting to emerge through a crack in the wall.

very serious proportions in the United States during the past generation.

There is one disease-producing Ascomycete which has been turned to advantage by man.* Ergot is a disease of rye and some other

---

* In another way man has often put pathogenic (disease-producing) organisms to work for him. When man's interests are seriously threatened by some insect or other animal pest that has recently invaded the region, he may deliberately introduce a pathogen which is known (elsewhere in the world) to be an effective natural enemy of the pest in question. At times this move has succeeded in initiating an epidemic which wipes out the pest or greatly reduces its

cereals in which the grain becomes enlarged and virtually supplanted by a hard mass of fungus mycelium that contains a unique chemical (Fig. 30). Though this chemical is quite dangerous to cattle and horses when they accidentally eat the ergot in uncontrolled quantities, the extract has a large medicinal value in controlling uterine contractions in women.

FIG. 30.—Ergot. Sketch of one "head" of an infected rye plant, showing replacement of several of the grains by the hard, compact, dark-pigmented mycelium of the fungus.

In addition to the parasites there are many saprophytic Ascomycetes of common occurrence. An Ascomycete is responsible for the blue mold that appears on stale bread, on preserves, and on old leather. An Ascomycete is responsible for the dark spots we see in Roquefort cheese and probably for the distinctive flavor of the cheese itself. The "cup-fungi" that we so often see growing on rotting logs in the forest also fall into this group. The "cup" itself is an elaborate reproductive branch put up by the mycelium, and the brightly colored lining of the cup is composed of thousands of asci standing compactly side by side (Fig. 31). Though the majority of edible fungi

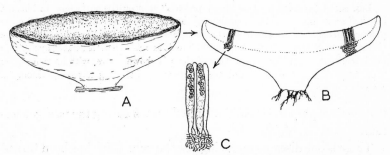

FIG. 31.—A cup fungus. A, as it might appear in nature; B, sectioned longitudinally to show relative position of the asci and the matted mycelium which composes the bulk of the tissue of the cup; C, an enlargement of the asci, interspersed with erect, sterile filaments.

numbers. Unfortunately there are also cases in which the pathogen has refused to "stay within bounds" but has gone on to attack other hosts which man wished to preserve.

are members of the Basidiomycete group, the Ascomycetes furnish
two edible forms that are usually regarded as rare delicacies. In the
"morel" the saprophytic mycelium, growing
on decaying vegetation at or near the soil surface, puts up an elaborate reproductive branch,
in which the pits of the honeycombed surface
are lined with a compact stand of finger-shaped
asci (Fig. 32). In the "truffle" the reproductive
branches remain so nearly buried in the litter
of grass and dead leaves that the French have
been obliged to employ trained pigs to scout
out their location.

Yeast is a saprophytic plant that is commonly classified among Ascomycetes by virtue of its
occasional production of an ascus-like sac of
spores. The classification is rather question-
able, particularly as this form lacks the mycelium that is characteristic of fungi in general. The body of the yeast organism consists

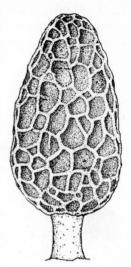

FIG. 32.—A morel

of a single (usually oval) cell, with the qualification that the cell,
under the conditions most favorable to growth, develops into short
chain-like colonies by the unique process of "budding." In ordinary

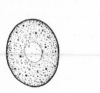

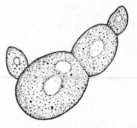

FIG. 33.—Yeast cells, highly magnified, showing the process of "budding," a form of
"vegetative multiplication."

cell division the two resulting daughter-cells are of essentially the
same size, and the individualities of the two are delimited with comparative suddenness. In the budding of the yeast, the prospective
daughter-cell first appears as a tiny "blister," pushing out from the
surface of the original cell, and gradually grows while still in intimate

cytoplasmic connection with its parent. A parent may put out more than one such bud, and the buds themselves, even before they are full sized, may start to produce secondary buds. Thus there results a colony in the form of a simple or branched chain, but the colony readily breaks apart into its component cells (Fig. 33).

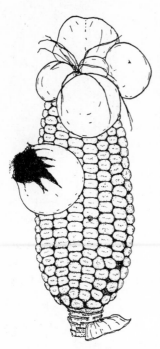

FIG. 34.—Eruption of "smut" pustules on ear of corn.

The economic importance of the yeasts arises from their remarkable capacity to conduct fermentation. Though most organisms can secure the energy from their food only with the assistance of oxygen in its free gaseous form, there are some, and most notably the yeasts, which can utilize their foods in the absence of free oxygen. Through this power of fermentation they can live, therefore, under conditions which would quickly suffocate most forms of plants and animals. Yeast, in its fermentation, breaks down sugar into alcohol and carbon dioxide. In the commercial production of alcoholic beverages and of pure alcohol itself, yeast is put to work on some form of sugar solution. In "raising" bread it is the other product, the carbon dioxide gas, which does the work.*

Just as Ascomycetes are characterized by the ascus, so the Basidiomycetes are characterized by the "basidium," which occupies the same place and plays the same rôle in the life-cycle. Sex acts, often even more obscure than those of the Ascomycetes, are indirectly responsible for the basidia, and the early development of the individual basidium runs parallel

---

* Some other organisms can also ferment sugar, and in these other cases the sugar may be broken down in other ways, yielding other products than alcohol. In the muscles of man himself there occurs a fermentation of sugar to lactic acid. This provides energy for muscular movement much more rapidly than if the muscles had to await the comparatively slow delivery of free oxygen by the blood stream.

with that of the ascus. But the final result is different, for the mature basidium commonly produces only four spores, and these are carried at the ends of little stalks. From this we recognize that the wheat rust fungus is a Basidiomycete and that the little mycelium which emerged from the "winter spore" is its basidium (Fig. 25).

The most conspicuous categories of pathogenic (disease-producing) Basidiomycetes are the smuts, the rusts, and the wood-destroyers.

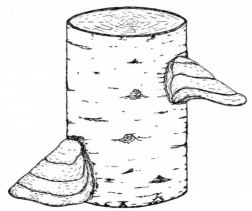

FIG. 35.—Two "brackets" emerging from the trunk of a host plant, within which the mycelium of the wood-destroying fungus has probably been present for many years. The tiny pores which cover the under surfaces of the brackets are the openings of cylindrical chambers which are lined with basidia.

The various smuts occur not only on cereals but on onions and some other garden vegetables as well. Extensive compact masses of black spores form in the grain or some other tissue, later to erupt in an unsightly mess (Fig. 34). In one smut, which is commonly known as "bunt" or "stinking smut" of wheat, the spore mass yields an unbelievably bad odor (the odor of trimethylamine). The spores referred to are the important ones in distributing the species but are not the product of the basidium, which occurs at another point in the life-cycle.

The rusts attack a wide range of plants. Not all, but many of them exploit alternate hosts, the two hosts never being very closely related forms. Man commonly attempts to combat the rust by exterminating whichever host is the less valuable.

The wood-destroyers, usually penetrating through a wound in the tree, may persist for decades within its host, until at last the weakened stem crashes in a wind storm. For reproduction the internal mycelium puts out the well-known "bracket," which carries millions of basidia with their spores (Fig. 35).

Basidiomycetes also provide a picturesque array of saprophytic forms, commonly referred to as the "fleshy fungi," among which we find certain distinctions between the "mushroom" group and the "puffball" group.

FIG. 36.—Mushroom. The mycelium ("spawn") ranges through the decaying material at or near the soil surface, and at last gives rise to the umbrella-like reproductive branches. The "button" shown at the right is the immature stage of the reproductive branch.

In the mushroom group the mycelium, ranging through the decaying vegetation at or near the soil surface, or through the tissues of a dead log or stump, puts forth reproductive branches which are enormous as compared with the vegetative mycelium itself. These branches are really very complex spore-bearing stalks, their "fleshy" substance being composed of a fairly compact arrangement of thousands of mycelial threads. In general contour they show variations on two conventional patterns, the "bracket" which extrudes from a log or stump, and the "umbrella" of the edible mushroom which arises from the ground. The fertile tissue of the reproductive branch is in the form of an extensive single layer of basidia, each of which yields the four (occasionally only two) spores. In the common edible mushroom and in many others this fertile layer coats the "gills" which hang in radial arrangement from the lower surface of the umbrella (Figs. 36, 37). In many forms the corresponding surface is flat, pierced by a vast number of tiny pores, the external openings of tiny cylindrical chambers which are lined by the layers of the basidia. The shape of the reproductive branches of some types is responsible for their name of "coral fungi"; here a good part of the outer surface is coated by the basidium layer. This is also true of the "spine fungi," where the spore-branch splits up rather irregularly and yields innumerable pendent spines (Fig. 38). The startlingly white

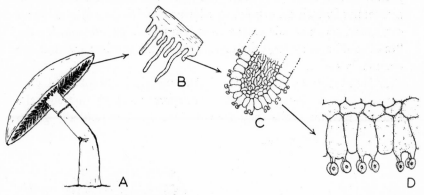

FIG. 37.—Spore production in the mushroom. *A*, the entire reproductive branch; *B*, longitudinal section through a few of the pendent gills; *C* and *D*, successive enlargements to show basidia and spores.

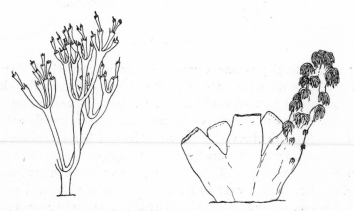

FIG. 38.—Sketch of "coral fungus" on the left, and of "spine fungus" on the right

color of the young spine fungus makes it an object of rare beauty, suggesting a winter scene from some Lilliputian forest.

The term "toadstool" carries no botanical significance, for the fungi referred to by the term do not constitute any natural grouping. Scattered here and there through the mushroom group there are a few forms—fortunately fewer than is popularly believed—which produce poisonous alkaloids. These substances are of no apparent value to the organisms themselves, but are merely the inevitable by-products of their life-processes. The writer knows of no sure criterion by means of which these poisonous forms may be distinguished in all cases, and recommends extreme caution in this connection, for the "toadstool" often closely imitates some of the mushrooms that are commonly eaten.

In the mushroom group of the fleshy fungi the layers of basidia are carried on exposed surfaces, so that the spores are in contact with the external medium while they are ripening. In the "puffball" group

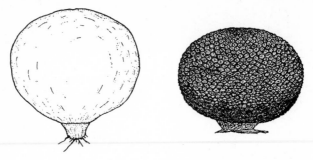

FIG. 39.—Puffballs

the spores develop within a general enclosure from which they are released only when they are fully ripe. The common puffball itself is the simplest expression of this device. Here the entire reproductive branch constitutes the enclosure within which basidia are carried on a network of mycelial threads (Fig. 39). The size of the individual puffball varies greatly (probably hinging on the conditions surrounding its development) some being known to reach several feet in diameter. If one were to canvass the entire plant and animal kingdoms for the organisms with the highest potential reproductive ratio, he would have to award first place to some of the puffballs, for, in some of the largest forms, the number of spores produced by a single individual must exceed 1,000,000,000,000.

The "earth star" is a puffball with a peculiar outer coat, which peels back in the form of a star to expose the "puffball proper" within. In the "bird's nest fungi" the pattern of the reproductive branches has to be seen to be believed. A surprisingly regular and trim nest of tissue contains a group of "eggs," each of which is actually a tiny puffball (Fig. 40). Another interesting member of the

FIG. 40.—Sketch of "earthstar" on the left, and of "bird's nests" on the right

group is popularly referred to by the inelegant title of "stinkhorn." Here a mushroom-like stalk is surmounted by an enlargement which at maturity exposes a honeycombed surface covered with innumerable black spores (Fig. 41). The spore mass exudes the odor of carrion which acts as a lure to the carrion flies. So far as the writer knows, the flies derive no benefit, but the fungus is effectively distributed by the spores that adhere to the bodies of the flies.

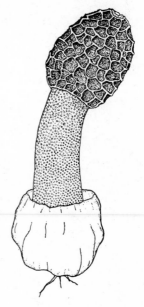

FIG. 41.—"Stinkhorn"

Proper classification of any fungus—or of any plant, for that matter—demands a knowledge of its complete life-cycle. If, for example, an ascus is present, the form will be put among the Ascomycetes; and then, on the basis of the detail of those structures which produce and surround the asci, as well as on the basis of the various spores and other features that may appear elsewhere in the life-cycle, the form will be placed in the proper order, family, genus, and species of the Ascomycete group.

There remain many fungi, however, in which as yet man knows only a part of the life-cycle, a part which includes no ascus or basidium, but only spore types which are less definitive. This is not surprising in view of the microscopic proportions of fungi and the obscure quarters in which they sometimes live. Yet botanists are obliged to provide some classification for these imperfectly known forms, if for no other reason, because they include many that are of great economic importance. He meets the situation by assigning them to a great group known as "fungi imperfecti." This he regards as a purely tentative assignment, for, as time passes, the missing stages of some of the life-cycles are revealed, so that steadily the members of the "fungi imperfecti" are transferred to the other groups, usually, as it turns out, to the Ascomycetes.

Many people, upon studying the dependent organisms and their various devices for exploiting living things, become genuinely depressed. To them it appears—particularly during their less vigorous and courageous moments—that a world in which success is built, in large part, upon a "ruthless" appropriation of the goods and destruction of the lives of other individuals is indeed an unhappy place in which to live. Since our nature is one in which matter and energy are never created, but only transformed, the building-up process could not continue (for long) save at the expense of tearing down something else. So destruction of a sort will have to be accepted as a "law of life," but to call it "ruthless" is to attribute to other organisms the concept of mercy which is peculiarly human.

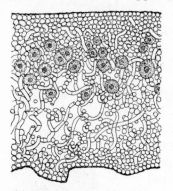

FIG. 42.—Microscopic view of longitudinal section through the body of a lichen, showing scattered cells of the alga component enmeshed in the mycelial threads of the fungus.

Some comfort, however, may be derived from the realization that destruction is by no means the only law of life. The needs of two organisms are often satisfied by a co-operation in which each is benefited and neither is damaged. A group of forms in which this is strikingly illustrated is the "lichens."

A lichen is a compound of an alga and a fungus. The alga, surrounded and kept more or less moist by several layers of fungus mycelium, can utilize the sunlight that falls upon spots that would be far too dry to support the alga alone (Fig. 42). The fungus also benefits by this partnership through sharing the food that the alga has manufactured, and does so without killing—and usually without even penetrating the cells of—its partner. Both can thus live in regions that can support neither one alone, and commonly in regions that can support no other form of life.

The alga component can be removed from the lichen and can be kept alive independently. In some cases this has also been possible with the fungus component by providing a suitable nutritive medium. Some lichens have been synthesized artificially by bringing together algae and fungi that had not previously been thus associated in nature. In natural lichens the alga component is most commonly a green alga of the *Pleurococcus* type, and the fungus component practically always an Ascomycete.

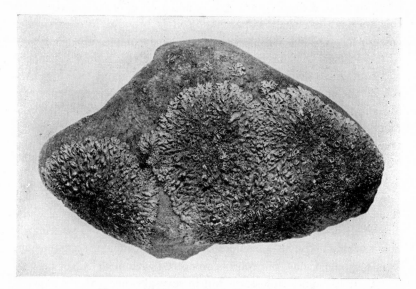

FIG. 43.—Photograph of lichens growing on rock

The alga, imprisoned as it is within the surrounding fungus, has no opportunity to distribute itself by itself. The fungus may produce spores superficially and scatter them, but the resulting mycelia must come to naught save in those extremely rare instances when they might encounter a new alga partner. Effective reproduction of the lichen is accomplished by little bud-like structures which include both alga and fungus elements, and which may be distributed by insects and other animals.

Some of the lichens are "professional pioneers." Repeatedly in the past, movements of the earth's crust have exposed great stretches of bare rock surface, thus creating dry, "sterile" areas, hopelessly incapable of supporting the ordinary forms of plant and animal life. But on these sterile rocky surfaces the lichens may make a start. Here they can eke out a living, for they have remarkable powers of resistance and dormancy to tide them over the long dry spells; and when rain comes they can soak it up like so much blotting paper and retain it to support a more "active" life for quite a period (Fig. 43).

In the course of time, acids which are by-products of the lichen's life-processes may gradually crumble the rock surface; and the lichen itself, dying and disintegrating, may add a bit of organic matter. In this way a tiny, shallow patch of soil is brought into existence, enough to provide a foothold for small hardy plants of other types. These, in turn, disintegrate the rock surface still more extensively and contribute their own dead bodies to the accumulation of material at the surface. More soil is thus produced, larger plants can come in, and there is thus launched a "plant succession" which culminates thousands of years later in a mighty forest, rooted in the several feet of soil which now cover the rock. Had it not been for the original lichen, this part of the earth's surface would have remained quite unavailable to man and beast.

Already we have encountered several groups of organisms which could not with certainty be assigned to either plant or animal kingdoms.* Still another such group is the slime molds.† The remarkable working body, or "plasmodium," of a slime mold is an extensive, formless, naked mass of protoplasm dotted with nuclei. Not only in its nakedness and formlessness but in several other features as well, this body is strongly suggestive of such simple animals as *Amoeba*. Under certain conditions the entire body moves, like an animated mass of jelly. The motion is characteristically amoeboid, for it is accomplished by pushing out irregular protrusions of protoplasm and having the rest of the body flow into them from behind. Like *Amoeba*, too, the slime mold moves in response to certain stimuli, such as

---

* This was true of the Flagellates and of the Volvocales. There are some biologists who would make the same statement of the bacteria, though the majority are inclined to recognize them as plants.

† Technically referred to as "Myxomycetes" by those who regard them as plants, and as "Mycetozoa" by those who regard them as animals.

light, retreating from the sunlight until it has at last moved into some shaded place. Again like *Amoeba*, the slime mold flows around organic particles and engulfs them into its body, where they are gradually digested and assimilated.

Such are the activities of the slime mold under those conditions which favor bodily activity. Under adverse conditions, however, it may act in a manner characteristic of both the lower animals and the lower plants. Forming a protective crust around itself, it will pass into a state of dormancy, in which it may remain for many years without losing its viability.

It is in reproductive features that the resemblance to plants is seen. Animals seldom form spores, and when they do the spores are not surrounded by cellulose cell walls. The slime molds, however, apparently quite devoid of any form of sexual reproduction, produce spores freely, and the spores have the characteristic cellulose walls of plant spores.* In some slime molds the entire plasmodium rounds up to act as a simple or multiple spore-case. In others the spore-cases are amazingly complex—very regular little objects, put up on long stalks, and producing numerous spores on the elaborate network that is developed within the wall of the spore-case.

The walled spores, released and distributed by air currents, show a remarkable behavior when they reach a suitable substratum. From the broken wall the naked, uninucleate mass of protoplasm escapes and slowly moves over the substratum like a tiny *Amoeba*.† After a time many of these amoeboid cells approach each other and coalesce into larger and larger groups, forming at last a plasmodium with its continuous cytoplasm and numerous nuclei.

Altogether one is quite impressed with the weird combination of simple and complex traits that appears in the slime molds, and the remarkable plasticity that characterizes their life-cycle. It is no

---

* That is, of those plant spores that are released in an air medium.

† In some slime molds the *Amoeba* form (or lack of form) and method of locomotion may shift for a time to that of a tiny flagellate, which swims through a liquid medium with its single flagellum. These two phases are reported to be freely interchangeable with a third, in which the cell "encysts," rounding up again and lying dormant for a time within a protective wall. Before plasmodium formation, however, the cells return to the *Amoeba* condition.

wonder that the biologist finds difficulty in placing them with certainty in either plant or animal kingdom. If they are plants, they might be regarded as fungi on the grounds that they lack chlorophyll, but it would probably be wiser to regard them as a group by themselves.

The majority of slime molds are saprophytes, which grow in moist places such as rich soil and decaying wood. A few, however, are aquatic and some are parasites. An example of this last produces the fairly common disease known as "club root of cabbage." Here the slime mold, which may persist for years in infected soil, enters the roots of its host and stimulates them to a disorderly overgrowth which results in an enlarged but malfunctioning development of these organs.

CHAPTER VIII

THE INVASION OF THE LAND

IN HIS classification the botanist divides the plant kingdom into four great divisions. Algae and fungi (along with bacteria and a few other groups of minor importance) constitute one such division, together being referred to as Thallophytes; and it is this Thallophyte division that is apparently the most primitive of the four.

At one time the Thallophytes must have been exclusively water plants. The algae apparently made a start in fresh water, for today we find practically all the members of the two more primitive groups (blue-greens and greens) living in that medium. Later on, a few lines of descent evidently invaded the oceans, for today we find that with very few exceptions, the members of the more specialized groups (brown and red algae) live their lives in salt water. In all probability the fungi also made a start in water. This is suggested not merely by the hypothesis that they are derived from algae but also by the fact that there are today many water-inhabiting types among the more primitive (alga-like) classes of fungi.

Hence one pictures the ancient earth as devoid of any vegetation on the exposed land surfaces—a barren, lifeless land, for, if there were no land plants, there could hardly have been land animals. At the same time, many of the fresh-water streams and ponds and the oceans themselves may have been teeming with life. Countless billions of simple alga plants, endowed with chlorophyll, were busily converting some of the simple materials which bathed them into food and protoplasm. Primitive animals, which may well have put in their appearance by this time, were doubtless preying upon the food-containing algae. And, preying upon both algae and animals, were water-inhabiting fungi and countless tiny bacteria.

This state of affairs may well have continued for millions of years, with few significant changes in the picture. And all this time there lay at the very doorstep the great domain of the land surface, a treasure-land for living things, with its vast wealth of unexploited energy and its tremendous variety of prospective habitats.

Inevitably a few of the water organisms were cast out onto the land by waves, stranded by falling tides, or by the seasonally receding waters of fresh-water ponds and streams. In practically all cases the result was death. Protoplasm had been adapted to a water medium; it was fundamentally of such a nature that its activity and continued existence depended absolutely upon the presence of water. Stranded on land, the alga would be bathed in air, so that its protoplasm would dry out and die.

Some few of these early adventurers, however, may have managed to maintain themselves on land and there to eke out a precarious existence. Today one encounters a very few types of green algae (e.g., *Pleurococcus*) and a few more of the blue-greens that live, not actually submerged in water but in places that are moist, or usually moist. Perhaps these few types have been successful (where most of their associates failed) through some quality of their walls or sheaths which resisted somewhat the drying influence of the air that bathed them. Perhaps it was because they had perfected the trick of dormancy, putting their protoplasm to sleep on very low water rations during times that were dry, and reawakening it when rain, dew, or moist air returned. In any event, these diminutive early invasions led very little further. They suggest the early Viking colonies in North America. Thoroughgoing conquest was to come much later, and was to be initiated along other lines.

Fungi (and bacteria) are today fairly successful land-living forms. Only a moment's consideration, however, tells us that the invasion of the land by fungi was an event (or better, a series of events) that occurred comparatively late in the history of life on the earth. Fungi are dependent forms, parasites and saprophytes. The parasites could not have invaded the land until host plants and host animals were already established there. The saprophytes could have taken the step only after there had been enough living organisms on land to leave a litter of organic débris.

It is quite conceivable, however, that the lichens effected an early invasion of the land surface that was successful along limited lines. We have no evidence to tell us when this occurred. In any event, it is clear that lichens have not given rise to descendant forms that are significantly higher than themselves. The conquest of the land surface that was to lead to our higher plants was initiated by still another group of invaders.

One of the four great divisions of the plant kingdom is known as the Bryophytes. The first part of the term is derived from *Bryum*, a well-known type of moss; the last part, "-phyte," merely means "plant." The Bryophytes, then, are the "moss plants," containing mosses and their relatives. As a matter of fact, there are just two main subdivisions of this great group, the "liverworts," which are the more primitive, and the mosses, which are the more specialized.

Apparently the Christopher Columbus of the plant kingdom was a liverwort. It was this invasion of the land that led to its ultimate conquest.

The type of body which we find among the simplest of liverworts today suggests the nature of its ancient ancestor which first emerged from the water. Clearly this ancestor was a green alga, and in all probability it was a type of green alga that possessed a plate-like body. True, most green algae possess single-celled or filamentous bodies, and this was doubtless as true in the past as it is today, for such bodies are well adapted to a water medium. By the same token, however, such bodies are poorly constructed for life in the air. The single-celled type would have its entire cell surface exposed to the drying influence of the atmosphere; and in a filament, whether it is branched or unbranched, the bulk of the surface of every cell would be exposed. But if the body were a plate of cells (one layer of cells in thickness), all the cells, save those at the very edges, would be pretty well protected by contact with their neighbors, and exposed only on the upper and lower surfaces. Presumably such an alga, stranded on the bank of a pond, would stand a better chance of surviving than would most of its relatives.

Of the first million green algae with plate-like bodies that were left out on muddy banks, perhaps all but one wilted and perished. This one may have survived in the new habitat for any one of a number of reasons, but very likely because it chanced to be better adapted to life in the new medium, because it happened to possess some feature which better enabled it to resist the drying influence of the atmosphere. What this particular feature was is a matter of conjecture. About all that we can be at all sure of is that three significant features were soon introduced in connection with the invasion of the land, for even the simplest of the surviving liverworts possess these three features.

Very likely our successful pioneer and his descendants eked out a precarious existence on the muddy banks for many generations. This was a period of stress, during which the majority perished and the only survivors were those which happened to possess the most favorable combination of characters. At the end of this period all of the final survivors emerged with three characters which we may call the primary adaptations to life on the land.

The nature and value of these three adaptive features are easily understood. One—probably the first one—was a compact body, i.e., a body several layers of cells thick. In such a body all the cells, save those in the superficial layers, are insulated against the evaporating influence of the air, so that the total surface area of the plant is reduced in proportion to its volume, and the "average exposure per cell" is considerably less than in a body one layer in thickness.

FIG. 44.—Sketch of the prostrate, ribbon-like body of the liverwort, *Marchantia*

A second adaptive feature was the prostrate habit. Those platelike green algae which we find today either float free or are, at best, attached to their substratum by one "corner" only. This might be expected where it is to the advantage of the organism to expose a maximum of surface to the surrounding medium. Of necessity, however, the early land invaders must have adopted a different habit. The ribbon-like bodies of liverworts lie flat in the mud and are attached to it all along their lower surfaces. This reduces the exposed surface by practically half and maintains effective contact with the only available source of water (Fig. 44).

The third primary adaptation was the introduction of the first real "tissue," a tissue of vital significance to all land-living descendants, as is the corresponding tissue to all land animals. A tissue is a group (or a number) of cells which possess the same characteristics and which differ therein from other cells of the body. The tissue in question was the "epidermis." In plants, epidermis refers to the one superficial cell layer which surrounds the entire body.

The cells of the epidermis are packed closely together, leaving no interstices between, and (in regions where the body surface is exposed to the air) the cell walls (particularly the outer cell walls) are thickened and impregnated with a water-proofing material. Thus the land plant is provided with a natural "slicker," but the value of this slicker is in preventing the exit of water rather than its entrance.*

An innovation, valuable as it may be in serving pressing current needs, usually proves dangerous if carried to an extreme. An epi-

---

* To speak of compact body, prostrate habit, and epidermis as the "primary adaptations" to life on the land is a typically academic procedure. For the purpose of organizing their own thoughts on the subject, and, even more, for the purpose of conveying those thoughts to others in a fairly concise manner, school teachers are prone to describe a situation or episode in nature or society in terms of points number 1, 2, 3, etc. They state explicitly or imply that these points are the "significant," "fundamental," or "primary" points in the causation or characterization of the situation or episode, other points being of subordinate value. Often this usage is quite justified on grounds of rigorous logic, for the points mentioned do indeed come first in the chronological or causative sense. In other cases, however, the situation or episode described is a complex of many interacting factors, each one indispensable in making the affair what it is. When the academician selects from such a complex only a few points, and designates them the "primary" points, his procedure is arbitrary rather than logical. This may be the result of his ignorance or of muddy thinking, but often it is a deliberate violation of logic in the interest of practical utility. In other words, the professor, recognizing ten indispensable points, may, under pressure of limited time or interest on the part of his audience, select only three and stress them as the important points. This teaching device, used again and again by practically all teachers, serves the interests of breadth rather than depth. For the most part, it well serves the interests of general students of the subject. The academician clears his conscience a bit by revealing his trick to those students who are interested in greater rigor and penetration.

For teaching purposes, then, compact body, prostrate habit, and epidermis are the primary adaptations to land life. Actually these adaptations must have been evolved gradually, and meantime there were evolving several other related adaptations which were doubtless quite as indispensable.

dermis is of value in keeping a land plant from drying out, but a perfect and complete epidermis would kill the plant just as surely by starvation (as well as suffocation). The process of food manufacture by green plants requires light as a source of energy, water and carbon dioxide as the raw materials. Algae, submerged in fairly shallow water, get their sunlight from above, and readily take in from the surrounding medium not only water itself but carbon dioxide, which is dissolved in the water. For a land plant, such as our pioneer liverwort, the situation is different. Sunlight readily penetrates the transparent epidermis to the chloroplast-containing cells beneath.* Water is soaked up (diffuses) from the mud beneath and passes to the chloroplasts. For a land plant, however, the small amount of carbon dioxide that might be dissolved in the soil water would provide an inadequate supply for the purposes of food manufacture. Instead, the carbon dioxide of the atmosphere must enter in the form of a gas. It is for this reason that a complete epidermis would prove fatal. For the epidermis is impermeable to other gases as well as to water vapor.

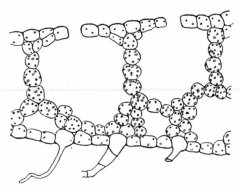

Fig. 45.—Longitudinal section through a very primitive liverwort body, showing simple breathing-pores in the upper epidermis, opening into simple air chambers beneath.

One finds, therefore, that the epidermis of every land plant is punctured by many microscopic "breathing-pores." In the simpler liverworts the breathing-pores are little more than irregular clefts in the skin (Fig. 45). Among some of the more specialized liverworts,

* The full "specialization" of epidermal cells was apparently a gradual process, and even among living forms we see various stages in the process. One criterion of this is the chloroplasts in the epidermis. Primitive liverworts have about as many chloroplasts (per cell volume) in their epidermal cells as in the cells beneath; more advanced liverworts have fewer. Plants of phyla higher than the Bryophytes characteristically lack chloroplasts in the epidermal layer.

however, the breathing-pore is a fairly elaborate device, surrounded
by a "chimney" arrangement of tiny epidermal cells, and leading
into a rather symmetrically designed "air chamber" beneath (Fig.
46). Plants of still higher phyla possess a fairly well standardized
type of breathing-pore in which the opening itself is surrounded by a

FIG. 46.—Longitudinal section through the body of the liverwort, *Marchantia*, showing one of the chimney-like breathing-pores opening into an air chamber. The filamentous extension into the air chamber of those cells which are best equipped with chloroplasts makes for a rapid supply of carbon dioxide to those cells that are using it the most.

pair of sausage-shaped cells (Fig. 47). By an interesting device these
two cells so change their contours as to increase or decrease the size of
the "breathing-pore" under different sets of conditions.

A biological principle that is illustrated by countless examples in
the plant kingdom, and by even more striking examples in the animal
kingdom, is as follows: Organisms which are usually surrounded on
all sides by the same conditions have homogeneous bodies; organ-

isms which usually encounter different sets of conditions on different sides are organized into different body regions, each region being adjusted to the conditions which it meets.

Among the highest types of plants and animals this specialization of body regions is quite complex and adapts the organism remarkably well to the general enterprise of "getting the most that is possible out of the environment." But these niceties of adjustment have not been worked out over night. They are the end-products* of an incredibly long series of evolutionary changes.

An illustration of the earlier and simpler stages in the differentiation of body regions is provided by the liverworts. Here the body shape and orientation is such as to create two distinct surfaces, upper and lower, which continuously encounter different sets of environmental conditions. The upper surface is coated with a protective and transparent epidermis,

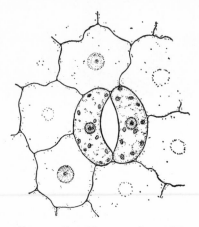

FIG. 47.—Breathing-pore of one of the higher plants as it might be seen by microscopic examination of the lower surface of a leaf.

save for the tiny breathing-pores which permit the entrance of gaseous carbon dioxide in the interests of food manufacture.

Beneath the breathing-pores the first few layers of cells are in a spongy rather than compact organization, i.e., a system of connected air channels permits the carbon dioxide to reach the surfaces of all the chloroplast-containing cells.

But the cells in the lowest few layers of the body are poorly situated for food manufacture. They are virtually "in the shade," for the cells above have caught the sunlight first, leaving only a tiny

---

* Today we might refer to them as "end-products." But evolution will go on in the future as it has in the past, and what today may appear to be an end-product of evolution may later prove to be only a transitional stage toward something still more advanced.

remnant of the energy-giving rays to filter through. One finds the cells of the lower layers in fairly compact arrangement. Their job is not food manufacture but water intake. The lowest layer of all is, technically, an "epidermis." On this part of the epidermis, however, cell walls are thin and are *not* impregnated with waterproofing material. It is to the advantage of the plant that water should pass easily through this surface.*

The intake of water through the lower surface of a liverwort will be proportional to the area of that surface. Here one encounters a biological principle that comes into play in many and various situations in the plant and animal kingdoms. Where the amount of the function depends upon the area of some exposed surface (and where it is to the advantage of the organism to increase the amount of that function) there will be a change, in the course of evolution, from a plain to an irregular surface. The absorbing area on the lower surface of a liverwort is greatly increased by extending some of the cells in long, hair-like processes that penetrate the soil (Fig. 46).

Here, then, one sees the beginnings of a differentiation into upper and lower body regions. The sharpness and the elaborateness of this differentiation is less among some (presumably the more primitive) than among other (presumably the more advanced) liverworts, and in no liverwort is the differentiation carried as far as among plants of still higher groups.

---

* With but few exceptions, land plants are unable to take in the water that might occasionally reach their exposed surfaces in the form of rain or dew. Their only source of water is the soil; where no water is present in the soil, as in a desert, plants are unable to live.

CHAPTER IX

REPRODUCTION AMONG LIVERWORTS

IN THE liverwort group as a whole, or "Hepaticae," as they are called, botanists recognize three subdivisions or "orders." (1) The "Marchantiales" possess ribbon-like bodies that are quite simple in their external outlines, but usually thicker (with more cell layers from top to bottom) and internally more complex (with comparatively elaborate arrangements of breathing-pores and air chambers, viz., Fig. 46) than those of the other orders. (2) The "Jungermanniales," or "leafy liverworts," have bodies that are comparatively thin and simple in internal structure, but complex in outline, for here the main axis of the body gives off a series of beautiful little leaf-like lobes. (3) The "Anthocerotales" have bodies that are simple both in outline and in structure but possess some significant peculiarities in their "sporophyte generation," as will be described toward the end of this chapter.

This is the general type of thing that has happened again and again in the evolution of plant and animal kingdoms. A line of pioneers, taking up a radically new mode of life, soon breaks up into several distinct lines of descent. In the subsequent evolution of each of these subordinate lines it is usually possible to recognize some peculiar evolutionary trend or "motif," some dominating character or combination of characters which is evolved step by step to reach its most exaggerated condition in the most specialized and chronologically the most recent products of the line.

It would be surprising indeed if the land habit, which brought the innovations that we have seen in the vegetative body, did not as well bring innovations in reproductive methods. Since the three orders of liverworts are fundamentally alike in their reproduction, we shall take up only the reproduction of the Marchantiales, adding at the end a brief comment on one peculiarity of the Anthocerotales.

One of the most conspicuous and best known of the liverworts is the genus *Marchantia*, from which the name of the corresponding

order is taken. In *Marchantia* one finds three methods of reproduction, three more or less complex life-cycles which happen to correspond (in more respects than meet the eye) to the three life-cycles of
the bread mold.

At one end of the ribbon-like body of *Marchantia* one always finds
a characteristic notch. It is a small group of actively dividing cells
right around this notch that is primarily responsible for growth of
the body. The notch is the growing-point, and moves forward as it
lays down new cells behind it. Occasionally the growing-point divides itself into two co-ordinate growing-points, each of which continues with the process of laying down new cells. This results in a
forking of the ribbon-like body into two small ribbons of equal magnitude (Fig. 44). Later each of these branches may fork again, and
so on, without any apparent limit. Under this system of growth the
youngest parts of the body are always at the front, nearest the notch,
the oldest parts being behind. Hence death "from old age"* occurs
first in the rearmost portion of the ribbon and moves forward from
that point. When death reaches a fork it effectively divides the
plant in two, for the two branches, no longer connected by living
tissue, continue to carry out an independent existence. Here, then, is
about the simplest conceivable method of "vegetative multiplication" or "fragmentation," corresponding to what we encountered in
the bread mold and to what occurs in innumerable other plants.

A second method of reproduction that occurs in *Marchantia* and in
some of the other liverworts is apparently a more refined and specialized method of vegetative multiplication. Here and there on the

---

* Death "from old age" is a vague expression which covers up a great deal of
ignorance, for death in general, and particularly the kind of death that appears
to come inevitably at a certain age regardless of the nature of the surrounding
conditions, is among the most obstinate of biological problems. One factor is
"auto-intoxication." The very life-activities of protoplasm yield by-products or
wastes which, if they accumulate beyond a certain point, have a damaging
effect on the protoplasm itself. Animals have "excretory systems" for the
elimination of these wastes, though the excretory systems are, unfortunately,
far from perfect. Plants have practically nothing of this sort. If excretion of
wastes were perfect, *perhaps* protoplasm might be immortal. But, whatever the
factors involved, it is unquestionably a fact that a living cell, unless it manages
to divide in the meantime, sooner or later dies in all cases ever observed by man.

upper surface of the body there appear dainty "cupules" about one-eighth of an inch in diameter. Within each of these little cups there develop a score or so of tiny disk-shaped bodies, each at the upper end of a very short stalk. These bodies, single-celled at the outset, are at maturity many-celled affairs a little smaller than the smallest pinheads. Each disk bears a pair of notches on opposite sides, so that the whole disk suggests an attached pair of "twin" parts; hence the name "gemmae" has been applied to the disks (Fig. 48). Breaking from their stalks, the gemmae are sometimes distributed by becom-

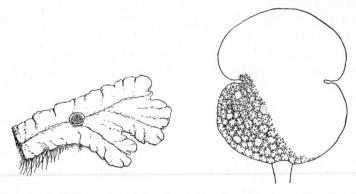

FIG. 48.—On the left, a "cupule" on the upper surface of the body of *Marchantia*, as one might see it under a hand lens. On the right, a microscopic view of a single "gemma," together with a portion of the stalk which carried it.

ing attached temporarily to the bodies of birds, insects, or other small animals. If a gemma reaches moist soil, it will proceed to grow, for each of its notches contains a growing-point. Theoretically it should develop into a double-ended *Marchantia* plant. Actually, one of the growing-points almost always gains an ascendancy over the other, so that the resulting *Marchantia* plant develops in only the one direction.*

---

* Botanists commonly think of methods of reproduction as falling into the following fundamental categories:
I. Asexual reproduction
   A. Vegetative multiplication, where the new individual is derived from a group of cells that becomes separated as a group (accidentally or systematically) from the parent-body. The title is justified by the fact that

The third type of life-cycle that appears in *Marchantia* is much more elaborate. At certain seasons some of the growing-points cease adding to the length of the body and produce instead vertical stalks surmounted by horizontal disks. Prior to this time all *Marchantia* plants have looked alike, but now it becomes evident that there are really two types of plants with respect to their potentialities. On some of the plants the stalked disks are comparatively flat and almost perfectly rounded, with only a series of slight indentations along their borders; but on other plants the corresponding indentations are much greater, so that the edges of the disks are broken up into a series of pendulous promontories. As we shall see, the former plants are male, the stalks they yield being "male gametophores," and the disks "antheridial disks"; the latter plants are female, their stalks "female gametophores," and their disks "archegonial disks."

A vertical section through an antheridial disk (Fig. 49) reveals several cavities within. In each cavity, and almost filling the cavity, is an "antheridium," or male sex organ. This is a many-celled affair, composed of a short sterile stalk and an enlarged upper portion. The outermost layer of this upper portion is merely a protective epidermis. Within is a solid mass of packed cubical cells, each of which, at maturity, produces two tiny ciliated sperms. With the maturing of

---

the group of cells is usually a functional part of the parent-body prior to the time of separation.

B. Spore reproduction, where the new individual is derived from one of many single cells that have been given off as single cells by the parent-body.

II. Sexual reproduction, where the new individual is derived from the fusion of two cells.

This classification may adequately dispose of the bulk of cases, but, like most arbitrary, man-made classifications, it may prove unsatisfactory when we attempt to apply it to certain cases. So often is it true in biology that the most reasonable interpretation of things is to be found not in their present-day constitution alone but in the light of their evolutionary history. The present-day constitution of the gemma of *Marchantia* suggests that it is a unit for vegetative multiplication, for it is indeed "a group of cells that becomes separated as a group from the parent-body." Historically, however, the gemma may well have been derived from the simple spore of some ancestor, with the innovation that the spore divided a few times before the moment of separation. (See also discussion of categories of reproduction that appears at the end of chap. v.)

the sperms, the wall of the antheridium breaks, and a small passageway between the antheridial chamber and the upper surface of the
disk opens. A drop of rain or a film of dew provides a medium
through which the sperms can swim out into the open.

Meantime "archegonia," the female sex organs, have been developing on the archegonial disks. In this case the sex organs are not
buried in cavities but hang down from the lower surface of the disk

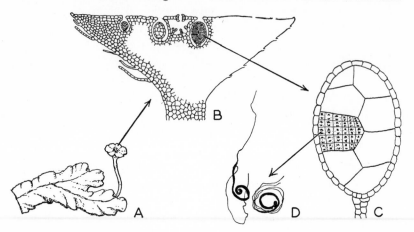

FIG. 49.—The sperm-producing structures in *Marchantia*. *A*, horizontal body of
male plant with erect male gametophore surmounted by antheridial disc; *B*, longitudinal section through antheridial disc, showing antheridia carried in chambers near the
upper surface; *C*, enlargement of a single antheridium, with stalk, epidermis, and mass
of sperm-producing cells; *D*, mature sperms, greatly enlarged.

(Fig. 50). The archegonium is also a many-celled structure, shaped
like an old-fashioned wine flask, with long, thin neck and swollen
base. The swollen base, which is attached to the archegonial disk,
consists of a protective layer (usually becoming more than one cell
in thickness as the archegonium matures) and one large, passive egg
within. The long, pendulous neck is perfectly sterile. In a young
archegonium this neck is solid tissue, but, by the time the egg is mature, a passageway opens up through the core of the neck, leading
from the free end of the neck up to the egg itself.

Since the sperm requires a medium of water for its movement, the
transfer of the sperm from the point of its origin to its functional
destination becomes a serious problem in *Marchantia*, where the sex

organs are produced upon different plants. As among many algae, the egg exudes a chemical which directs the sperms in their movements, but only a very rare set of conditions would make it possible for this chemical to diffuse through a continuous medium of water over to the antheridial disks. Actually, fertilization fails very often in this genus. The conditions favoring fertilization are a dense growth of mixed male and female plants (so that the two types of

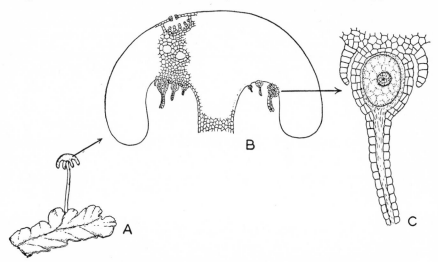

FIG. 50.—The egg-producing structures in *Marchantia*. *A*, horizontal body of female plant with erect female gametophore surmounted by archegonial disc; *B*, longitudinal section through archegonial disc, showing archegonia hanging from the lower surface; *C*, enlargement of the flask-shaped archegonium containing its single egg.

gametophores form a thick miniature forest)and a drenching rain. At such a time some sperms may indeed find a continuous water passage to the eggs, but probably still more of them reach their destination by being splashed over from antheridial to archegonial disk.

The resulting zygote germinates immediately. Its product is not another plant of the type on which it was produced, but a plant of the "alternate generation." In liverworts, and in all higher plants, sexual reproduction is consistently tied up with a life-cycle that is characterized by the "alternation of generations." The plant (i.e., male and female plants) which has already been described, constitutes the "gametophyte" (gamete-producing plant). The gametophyte generation produces gametes which yield a zygote. This zygote develops into the "sporophyte" (spore-producing plant).

In *Marchantia* the gametophyte is the green, independent generation of the two. The sporophyte, practically devoid of chlorophyll, develops parasitically at the expense of the gametophyte. The first cell of the sporophyte generation is the zygote itself. The zygote, re-

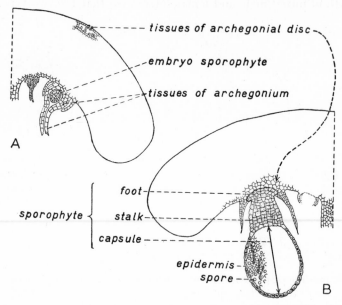

FIG. 51.—The sporophyte generation of *Marchantia*. *A*, young embryo sporophyte retained within the swollen base of the old archegonium; *B*, mature sporophyte which has burst through the old archegonium and is now anchored by its foot in the tissues of the archegonial disc.

maining within the swollen base of the pendent archegonium, divides repeatedly to form a mass of cells. These cells soon become organized into three body regions: (1) a "foot," which pushes its way back into the tissues of the archegonial disk, anchors the developing sporophyte, and secures food for it from its host; (2) a "stalk" just below the foot, short at first, but elongating just before the maturity of the sporophyte; and (3) at the lower end, a "capsule," which becomes a spheroidal body with an inclosing epidermis and a mass of spores (together with a few specialized sterile cells) within (Fig. 51).

During its early stages of development the sporophyte is retained in the swollen base of the old archegonium, which for a time enlarges to keep pace with the growth of the embryo plant within. Sooner or later, of course, the archegonium wall is ruptured, and shortly thereafter the epidermis of the capsule breaks, releasing a shower of ripe spores. (The same archegonial disk commonly carries quite a number of sporophytes in various stages of development.)

As in the air-living fungi, these spores, instead of being naked and ciliated, are covered with walls that resist desiccation. Clearly they are adapted for distribution by currents of air, and the average extent of their distribution will depend on the height from which they are dropped. In this connection one notes an unusually awkward form of adaptation in *Marchantia*. To facilitate distribution of spores, *Marchantia* elongates its gametophores. In a way this effects the desired result, but at the expense of putting the sex organs in such a position as to make fertilization difficult. Living organisms are, in general, remarkably efficient bits of machinery, but not uncommonly one encounters adjustments that are seemingly quite clumsy. One must conclude that the capacities for adaptation

Fig. 52.—*Anthoceros.* Three sporophytes arising from the horizontal gametophyte. In each case the sporophyte foot is invisible, being bedded back in the tissues of the gametophyte. The little "collar" around the base of each sporophyte consists of gametophyte tissue.

by living organisms in the course of their evolution are by no means infinite, but are actually restricted within rather narrow limits that are set by the structural peculiarities of their ancestors.

Reaching moist ground, the spore germinates to produce the gametophyte generation. Though all spores are quite alike in appearance, and in the apparent characteristics of their immediate products, it is obvious that the spores are of two types with respect to their potentialities, for some of the young gametophytes develop into males and the others into females.

Among other members of the order Marchantiales and among the Jungermanniales, reproductive methods are essentially the same as in *Marchantia*. Only in the Anthocerotales do we encounter a sig-

nificant difference. In the genus *Anthoceros* the gametophyte generation is quite simple in both form and structure. The sex organs, instead of being borne up on elaborate gametophores, are simply produced in little groves on the upper side of the gametophyte. Both
antheridium and archegonium are similar in structure to those of
Marchantia, but the latter has its neck pointing up rather than down.
After fertilization—a much simpler matter than in *Marchantia*—the
zygote immediately develops the sporophyte. Necessarily this sporophyte grows upward from the gametophyte rather than downward

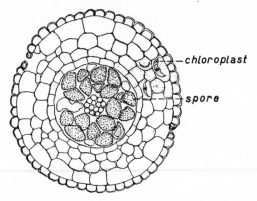

FIG. 53.—Enlarged cross-section of the green sporophyte of *Anthoceros*

from the archegonial disk. A recognizable "foot" pushes back into
the gametophyte tissue, but the rest of the sporophyte, instead of
being differentiated into "stalk" and "capsule," consists merely of a
cylinder of tissue, about the size of a small grass blade, which rises
vertically into the air (Fig. 52).

In cross-section, the sporophyte of *Anthoceros* reveals a series of
tissues, arranged concentrically: on the outside, an epidermis; next
a zone of sterile cells, several cell layers in thickness; next a zone of
"sporogenous" (spore-producing) cells; and on the inside, a core of
tiny, sterile cells (Fig. 53). The sporogenous cells at the upper end
are the first to yield mature spores, and this ripening process occurs
progressively from the top to the bottom of the sporophyte. As fast
as the spores are ripe, the sterile regions outside them split and peel
back, somewhat like one might peel a banana, releasing the spores

for distribution. Thus the spores are distributed periodically over quite a period of time.

Perhaps the most significant feature of the *Anthoceros* sporophyte, however, is connected with the sterile zone which lies just inside the epidermis. Each cell in this zone contains a large chloroplast. Thus the *Anthoceros* sporophyte is green and conducts food manufacture. (The epidermis is pierced by occasional breathing-pores.) Through the instrumentality of its foot, the sporophyte still depends on the gametophyte for water and soil salts, but it has taken the most significant step in shaking off its dependency and becoming a self-sufficient generation. As we shall see later, the significant progress of plants of the higher groups was based upon the attainment of an independent sporophyte generation. Of all Bryophytes known, *Anthoceros* most strongly suggests the condition of the ancestor from which the higher groups descended.

CHAPTER X

MOSSES AND THE ALTERNATION OF
GENERATIONS

THOUGH liverworts developed the ability to live on land they never got very far away from the water. Today we find them growing only in very moist places, on the banks of ponds or streams, on the rocks of shaded ravines, or around the bases of the trees of dense forests. A few liverworts have, in fact, reverted to life in the water itself, where they grow floating on the surface.

Though the exact relationships are by no means clear, it seems probable that mosses are the modified descendants of some of the earlier liverworts. Mosses, too, are usually found only in rather moist places, though some few have been evolved which can withstand quite a bit of desiccation, and can do a fairly good job of maintaining themselves in situations where moisture arrives only occasionally.

Almost everyone has encountered mosses frequently enough to hold a mental picture of them as tiny plants with numerous delicate "leaves." But the things that we ordinarily see and recognize as moss plants are in reality nothing but branches of the primary body. The primary moss body which emerges from a spore is nothing more pretentious than a branching, green filament, resembling the body of a green alga. This body grows horizontally on the surface of moist soil, soaking up water and conducting food manufacture by means of its chloroplasts. Altogether it is so tiny that it usually escapes the notice of the casual observer.

After a time, however, this horizontal, filamentous body gives off one or more vertical branches. The branch is much larger than the primary body, being many-celled in caliber, and it is covered, throughout its length, with a series of tiny, leaf-like structures. (Technically these are not "true leaves," though they resemble the true leaves of higher plants not only in their superficial appearance but in their function.) It is this erect leafy branch that constitutes the moss plant of popular experience (Fig. 54).

The growing-point at the upper end continues to add to the length of the erect branch and to give off more and more "leaves" until the total size has been attained that is characteristic of the species. (This may be anywhere from a fraction of an inch to a few inches in height.) At this stage the growing-point, instead of adding

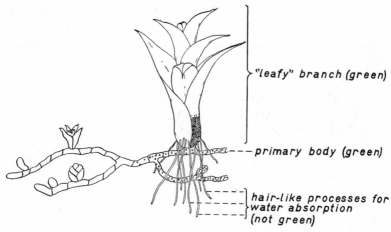

FIG. 54.—Diagram to show the relationship of the primary filamentous body of a moss and the "leafy" branch which arises from it. The filamentous body is seldom noticed by the casual observer, for it is relatively even smaller than the diagram indicates and is stretched out horizontally on the soil surface where it is often pretty well concealed by the soil particles.

further to the length of the vegetative body, devotes itself to the construction of the reproductive organs.

From the very tip of the axis there will arise several tiny, club-shaped structures. Examination under the microscope will reveal that each of these structures is a many-celled affair. The basal portion is merely a short stalk composed of several sterile cells, i.e., cells which are not destined later to yield reproductive cells. This short stalk is surmounted by a swollen and roughly cigar-shaped region, which consists of a single, sterile, epidermal layer on the outside and a mass of cubical, "spermatogenous" cells on the inside. As the name implies, the spermatogenous cells are destined to produce the sperms, each cell producing two of the very tiny, ciliated male gametes (Fig. 55). When the sperms are fully developed the sterile epidermal wall breaks, so that the sperms can escape and swim about in a tiny drop of dew or rain that may be present at the tip of the leafy branch.

The sperm-producing structures are commonly referred to as male "sex organs." It should be noted that these sex organs are

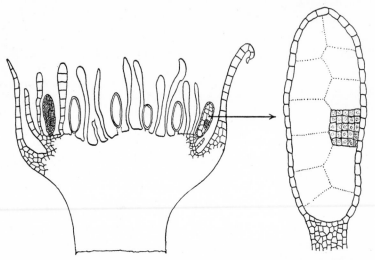

FIG. 55.—The male reproductive structures of the moss: at the left, a longitudinal section through the tip of a "leafy" branch, showing male sex organs interspersed with sterile hairs; at the right, an enlargement of a single male sex organ, showing the sterile stalk and epidermis and the fertile "spermatogenous" cells within.

many-celled. This is prevailingly true of the sex organs of Bryophytes. With but few exceptions, the sex organs of Thallophytes are single-celled; so that we have here one of the best of distinctions between the two great groups. The significance of the distinction is really connected with the land habit of the Bryophytes, for the many-celled condition is due (in large part) to the protective epidermal layer around the gamete-producing cells.

Female sex organs are also produced at the upper ends of the leafy branches. In some species of moss the two kinds of sex organs grow mixed together at the end of the same leafy branch, while in other species male and female organs are produced on separate plants. The female sex organ is also many-celled. Roughly it is the shape of an old-fashioned wine bottle with the swollen base below and the thin neck extending above. The base (at maturity) consists of one large, passive egg, surrounded by one or more layers of sterile epidermal cells. The neck above is quite sterile (Fig. 56). When the egg is ready to be fertilized, a passageway leading down to it is developed by the disintegration of a single column of cells which had formed the core of the neck, and down this passageway the sperm can swim, directed by a chemical which is exuded from the egg.*

---

* One is impressed repeatedly with the remarkably efficient manner in which the parts of living organisms are adjusted to the functions which they perform. But the student of biology encounters many exceptions to this rule. Structures are present because they have been inherited from ancestral types, and not all of these heritages are useful or efficient. In the present connection one may note two such heritages.

For a submerged plant, the ciliated sperm represents a device that is com-paratively efficient in bringing about fertilization. For the land-living Bryo-phytes, however, it would have been more useful to have a sperm that could move through the air from the point of its production to its normal destina-tion. But it was only many millions of years later that land plants were evolved which possessed such a device. Bryophytes still have the ciliated sperm of their alga ancestors. Hence fertilization is a possibility only when a continuous film of water is present through which the sperms may swim from the male sex organ to the egg. In those species in which male and female sex organs are carried together at the end of the same leafy branch, fertilization is not particularly difficult, for a tiny drop of dew provides ample water, and the chemical exuded from the egg directs the sperm in its passage. But in the other species, where male and female sex organs are carried on separate plants, fertilization is so difficult that it often fails completely. Apparently it is possible only where the two types of plants are growing close together. Under such conditions con-tinuous water passageways are often provided. Particularly is this true at times of rain, when the splashing effect may carry a few sperms directly from one plant to another.

Another inefficient heritage is the long neck of the female sex organ. Protec-tion for the developing egg is, of course, of vital importance, but this function is amply fulfilled by the layers of sterile cells immediately around the egg. There appears to be no value whatsoever in the long neck, which obviously makes it more difficult for the sperm to reach the egg than if a simple aperture had been provided. Probably this superfluous structure is a remnant of something that was useful in an ancient ancestor, of which the female sex organ may have con-tained a whole string of eggs rather than merely one.

The zygote that results from fertilization starts at once to produce the next generation. This new generation, however, is very different from the one which preceded it, for the life-cycle of the moss is characterized by an "alternation of generations." Prostrate filament and leafy branch are together spoken of as the "gametophyte genera-

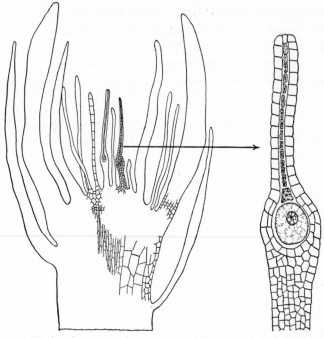

FIG. 56.—The female reproductive structures of the moss: at the left, a longitudinal section through the tip of a "leafy" branch, showing female sex organs interspersed with sterile hairs; at the right an enlargement of a single female sex organ, carrying its single egg.

tion" (or merely as "the gametophyte"). Gametophyte means "gamete-producing plant." The other generation which regularly alternates with the gametophyte in the life-cycle is known as the "sporophyte," meaning "spore-producing plant."

Among the Bryophytes the gametophyte is regularly a green, independent plant. The sporophyte, on the other hand, though it may for a time contain a few chloroplasts, is incapable of supporting itself. Instead it must live as a parasite at the expense of the gametophyte.

The young sporophyte of the moss, in the earliest increment of its growth, pushes a bulbous "foot" down into the tissues of the leafy branch, thus providing anchorage and a means of absorbing food for the mature sporophyte that is to come. With these basal connections well established, the developing sporophyte next enters into a period of rapid elongation, producing an upright "stalk" that is usually a little longer than the leafy branch that carries it. Clearly the function of the stalk is to raise to a greater height the spore-producing structures, thus facilitating the distribution of the ripe spores by chance air currents. As the stalk elongates, its tip region begins to swell, culminating at last in the production of the "capsule." This is the most complex structure that is to be found among Bryophytes. Its complexities appear to be devoted to the production of thousands of tiny spores, and to the release of these spores into the air when they are ripe (Fig. 57).

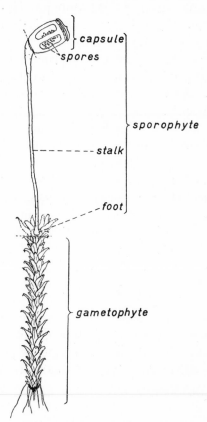

FIG. 57.—Diagram of the parasitic sporophyte of the moss, embedded in the leafy branch of the gametophyte.

The random distribution of these spores by air currents may bring some of them to rest on moist soil. These few fortunate spores, and only these few, succeed in producing the green, branching filaments of the next gametophyte generation. Thus sexual (gametophyte) and sexless (sporophyte) generations regularly alternate in the life-cycle of the moss.

In ascending the family tree of the plant kingdom it is in the Bryophytes that one first encounters this "alternation of generations" in a pronounced, easily recognizable form, with conspicuous differences between the green, independent gametophyte and the parasitic sporophyte. Actually the trained botanist is able to recognize the elements of this same life-cycle, with its alternating phases, even among a good many of the Thallophytes. Hence one cannot truly say that this feature provides a sharp differentiation between the two great groups. But it is the Bryophytes that provide us with the first obvious manifestations of the alternation of generations. As we shall see, the same fundamental type of life-cycle is retained, with modifications of some of the details, among the still higher plants, which were probably (but not certainly) evolved from Bryophyte ancestors.

Among the many sub-groups of the mosses there is one that deserves special mention. This is the group of the "bog mosses." Though these forms possess certain structural peculiarities, it is not difficult to recognize the correspondence between their structures and those of the other mosses. It is not their structure, however, that here commends them to our attention, but rather their habitat and their interesting action on that habitat.

Geological events, notably the action of glaciers, have left much of our country spotted with small ponds and lakes. It is along the shores of such small, quiet bodies of water that the bog mosses take up an abode. Starting at the very water-line, these little plants, by copious branching and vegetative multiplication, produce an ever widening mat of vegetation, which steadily pushes inward over the surface of the pond. In time the pond is completely roofed over by a fairly thick layer of the material, and is thus converted into a "quaking bog." If one walks out onto such a surface, it gives under his feet in a disquieting way, and the vibration spreads in all directions, since the surface layer is cushioned on water.

At a still later stage the pond is practically eliminated, being filled up completely by débris dropped from the bog moss, and by the older moss bodies themselves, which are constantly forced downward as new branches are added at the upper surface. This action paves the way for a succession of larger plants, whose root action and dying bodies gradually convert what was once a body of water into an area of rich soil. Our bog mosses have, thus, been the pioneers in a "plant succession" which culminates in the production of a forest of lofty trees.

CHAPTER XI

FERNS AND THE INDEPENDENT
SPOROPHYTE

IN HUMAN history the center of the stage has been occupied not by one race or nation continuously, but by a succession of different races and nations. One group of men, after some critical maneuver, occupies a strategic position, and may then strengthen that position over a period of time until it comes to dominate some region of the earth's surface. But this dominance does not persist: a different group supersedes the first. Furthermore, the new group, when we first catch sight of it, is usually more primitive than its predecessor, judged in terms of its entire culture or degree of civilization. Yet the newcomer possesses a certain "freshness" or "energy of youth" that enables it to build a new and somewhat different civilization. More often than not, the new civilization at last surpasses the old, so that over a long period of time progress inevitably results.

In the history of the plant and animal kingdoms a corresponding phenomenon is manifest. Though underlying causes may be somewhat different in the two situations, the effects are strikingly similar. One group of living organisms, after some critical evolutionary change, passes into a period of progress and conquest, based upon its ability to compete successfully with its neighbors or to enter new territory as a pioneer. This period culminates in the production of a considerable array of related types, the group as a whole having attained a greater complexity of structure and behavior than had ever been attained before. The new group is thus the dominant group of the times.

But this dominance does not persist. In the course of time a still newer group pushes itself into the center of the stage. The newer group, when we first notice it, is more primitive than its predecessor in relative complexity of structure and behavior. Yet it possesses a capacity for evolution along certain new lines. By exercising this capacity it becomes at last the dominant group of the times, represent-

ed now by a considerable array of related types which establish a
new high point in complexity of structure and behavior.

In short, the history of life on earth, like the far briefer history of
man himself, contains several major acts or chapters, in each of
which the same characteristic sequence of events is repeated. Repeatedly the relatively simple and primitive evolves into the more
complex and advanced. Repeatedly a comparatively humble and
distant relative of the earlier dominant group develops into the
dominant group of the moment. Repeatedly a somewhat new form
of progress supersedes and usually transcends an older form. Hence,
at any moment, the direct ancestors of those that are to dominate in
the future are likely to be found not in high places but in much less
conspicuous and less impressive positions.

In the history of the plant kingdom the Bryophytes, while by no
means exterminated, were superseded by another great group, the
Pteridophytes, which came to dominate the earth's surface to a degree that had never been attained by the Bryophytes. Though apparently the Pteridophytes descended from Bryophyte ancestors, it
was not from the relatively complex and successful mosses but from
one of the simplest and least conspicuous of the subgroups of liverworts.

Through the Bryophyte phylum as a whole, the gametophyte was
the dominant, green, independent generation. In their several subdivisions the Bryophytes worked out quite a number of variations on
this underlying "motif" of independent gametophyte and dependent
sporophyte. Some advances were made, but apparently there were
limitations to the amount of progress that could be based on such a
plan. The big plants of the higher phyla were to be made possible by
shifting the emphasis to the other generation.

Today there is one little group* of the liverworts in which the
sporophyte contains enough chloroplasts to be self-supporting. Since
it is still bedded by means of its foot in the tissues of the gametophyte, this sporophyte retains a certain measure of dependence, taking in its water and soil salts indirectly through the tissues of the
gametophyte. If it could take the additional step of rooting itself
directly in the soil this sporophyte would be independent.

---

* The Anthocerotales; see chap. ix.

There is reason to believe that liverworts of this type existed in the past, and that from this source at least one line of descent did indeed develop direct soil connections for its sporophyte generation. At first, no doubt, these independent sporophytes were puny affairs, and may have continued to be a very inconspicuous part of the plant population for millions of years. Step by step, however, they made evolutionary progress, steadily increasing the efficiency and the size of their sporophyte generation. Thus there emerged a new great division of the plant kingdom. As time went on these newly evolved Pteridophytes branched out into numerous lines of descent, getting larger all the time, until at last many of them took the form of trees which towered over their Bryophyte contemporaries and formed by far the most conspicuous part of the land vegetation.

The large-sized sporophyte, which rose to a considerable height from the soil surface, was made possible by a new type of organization, together with the introduction of certain new tissues. The body as a whole was organized into the three general body regions of root, stem, and leaf.* The roots provided for anchorage and the intake of water and soil salts; the stem provided the main axis on which the leaves were displayed in the sunlight; and the leaves conducted food manufacture for the benefit of the plant as a whole.

This in itself was not enough. More than ever before, provision had to be made to retain the water that was already in the plant, to secure a copious and fairly continuous supply of new water, and to transport this new water fairly rapidly to the uppermost parts of the plant. Retention of water was improved by the perfection of an epidermis which, while not differing in its general character, was somewhat more efficient than that of the Bryo-

---

* The student might be tempted to interpret these as the evolutionary products of the foot, stalk, and capsule of Bryophyte sporophytes. Such an interpretation would be rather unsound. The root system does indeed have some functional correspondence with the foot, as does the stem with the stalk, but it is difficult to see any significant connection between leaves, which are organs for the manufacture of food, and the capsule, which is an organ for the production and distribution of spores. As a matter of fact, if our interpretation is correct, the type of liverwort from which Pteridophytes descended had a sporophyte with no recognizable capsule, the spores being produced by certain tissues that were carried throughout the length of the stalk.

phytes. The increased supply of water was made possible by a large root system, which, through its many ramifications, presented a tremendous absorbing surface, and penetrated the soil to a greater depth than had any of the absorbing hairs of the Bryophytes. Rapid transport of water to the upper regions of the sporophyte body was made possible by a new tissue which consisted of a great many long, hollow tubes, joined end to end. By introducing these new specialized tubes, or "vessels," Pteridophytes became the first of the "vascular" plants. And vascular plants (which include not only the Pteridophytes but also the plants of the fourth and last great group) have been the only plants that have ever raised their bodies to any substantial height from the surface of the ground.

This vascular tissue which provides the highly significant conducting tubes is familiar to all of us as "wood." It is interesting that a tissue which has meant so much in plant history has also played a tremendous part in human history. Further details of the new, independent, woody, sporophyte body will be presented in a later chapter in terms of the type of plant which is much more familiar to the student than is the Pteridophyte.

By virtue of the new body organization and the new conducting tissue, the majority of Pteridophytes far outstripped the Bryophytes in the matter of size, and became the dominating land vegetation for a period of several millions of years. Landscapes came to bear a gross resemblance to those of modern times; vast forests sprang up, composed of tree Pteridophytes, some with stems 100 feet in height.

These plants were partly responsible for the "coal measures." The bodies of many of these large Pteridophytes, instead of disintegrating when they died, were preserved in swamps, gradually incorporated into the sedimentary rocks that were in the process of formation, and transformed at last into the coal which man, millions of years later, learned to put to his own uses.

In those early days Pteridophyte forests stretched over a wide range of latitude. Today such things are limited strictly to the tropics, where tree ferns, stretching forty feet or more in the air, are not an uncommon sight. But the Pteridophytes of temperate regions have been rather thoroughly superseded by a still more modern group, and the few that are left are indeed puny remnants of what was once a mighty array of plants.

Three major subdivisions of Pteridophytes are recognized—ferns, horsetails, and club mosses. Of these the ferns are decidedly the largest, in terms of number of species, and the most commonly encountered.

The typical fern of temperate regions has an underground stem, which in some ferns is in a "stubby," compact arrangement, and in others stretches out horizontally to a considerable length. Numerous small roots stretch downward from this underground stem. The only parts which extend above ground are the leaves. Fern leaves are characteristically large and compound, being divided (and often further subdivided several times) into many leaflets. The compoundness of the leaf, by itself, is not a very serviceable identification, for there are quite a number of higher plants which also possess compound leaves. There are, however, two characteristics of fern leaves that are fairly distinctive. 1. When the young fern leaf, put forth by the underground stem, first emerges from the soil, it is coiled up at the tip into a crozier-like formation. As development proceeds the leaf gradually uncoils until it has straightened out completely. 2. The smaller veins in a fern leaf exhibit a forked branching, i.e., the vein of larger magnitude splits into two equal veins of the next smaller magnitude. This contrasts sharply with the venation of higher plants, where the vein of larger magnitude gives off, along its length, a whole series of veins of the next smaller magnitude (Fig. 58).

Roots, stem, and leaves together constitute the sporophyte generation of the fern. Clearly this is an independent generation, capable of manufacturing its own food and procuring its own water and soil salts. As reproductive units it must, of course, yield spores. These are produced not in a single, specialized capsule but in a manner that is unique with ferns.

Even after the fern leaf has attained its maximum size it shows, for a time, no traces of the spores, but devotes its tissues completely to the manufacture of food. Oncoming age, however, apparently stimulates the under surface of the leaf to burst forth rather suddenly with a large crop of spore-producing structures. Doubtless the reader himself has noticed these numerous, small, brown or black patches scattered over the lower surfaces of the older fern leaves. These patches are so small as to reveal very little to the naked eye. The microscope, however, or even a good hand-lens, will show that each patch contains a score or so of tiny spore-cases. Each of the spore-

FIG. 58.—The sporophyte generation of the fern, with its underground stem and roots, and its compound, aerial leaves.

cases is carried on a long, thin stalk. In the earlier stages of development the patch of young spore-cases is covered by a protective flap of sterile tissue, on the principle of a pocket or an inverted umbrella (dependent on the type of fern—though there are some ferns that completely lack these protective flaps). As the spore-cases enlarge and ripen, the protective flap is pushed out and eventually sloughed off (Fig. 59).

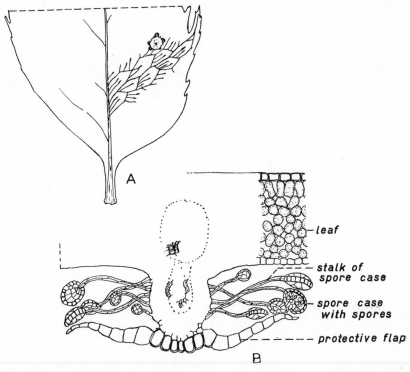

FIG. 59.—*A*, enlarged portion of lower surface of fern leaflet, as one might see it with a hand-lens, showing one of the patches of spore-cases covered by its umbrella-like (or shield-like) flap. *B*, longitudinal section cut through the leaflet at the point where one of these patches of spore-cases is carried.

Within a single layer of sterile epidermal cells each spore-case carries a group of spores, usually sixty-four in number. Mature spores are liberated, not by a simple rupture of the epidermal covering but by an active movement which serves to discharge them forcibly, on the principle of the catapult (Fig. 60).

Since a single spore-case contains sixty-four spores, since there are scores of spore-cases in a single patch, since there are hundreds or even thousands of these patches on the under surface of a large fern leaf and several leaves to a plant—the total annual output of spores from a single sporophyte may run way up into the millions. Here,

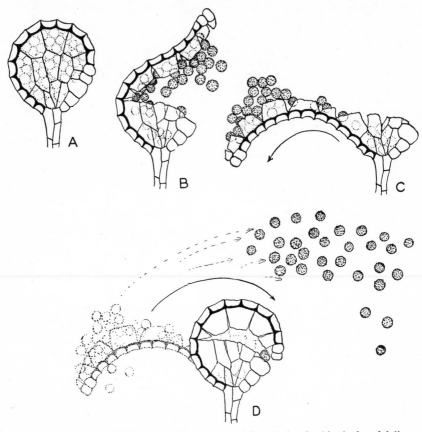

Fig. 60.—Series of diagrams to show sequence of events involved in the forceful discharge of spores by the spore-case of a typical fern. A single line of heavy-walled epidermal cells stretches almost completely around the spore-case, forming a spring-like structure. In response to the drying influence of the atmosphere this "spring" is slowly bent backward, and then very suddenly snaps forward again into its original position of rest.

then, is another example of a tremendously high potential reproductive ratio coupled with a high mortality; for it is only the tiniest fraction of the spores that reach conditions suitable for their developing any further.

The gametophyte that develops from the fern spore suggests the gametophyte body of a liverwort. Small (seldom over half an inch in diameter), flat, green, and anchored to the ground by a number of tiny, hair-like protrusions, it develops through the activity of a

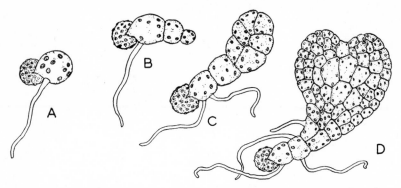

FIG. 61.—Series of early stages in the development of the fern gametophyte from the spore.

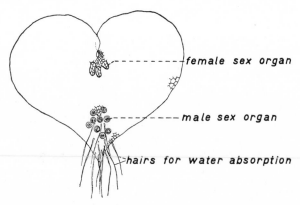

FIG. 62.—Diagram of under surface of mature fern gametophyte

growing-point in the base of a notch. Since forward growth does not continue indefinitely, the mature fern gametophyte lacks the ribbon shape of the liverwort, but culminates its existence with a heart-shaped contour (Fig. 61).

On the under surface of the gametophyte, next the moist soil, there develop the two types of sex organs (Fig. 62). These are essentially like the sex organs of Bryophytes with about the modifications in detail that one might expect from the fact that there is but little free space in which they may develop. The numerous male sex organs have negligibly short stalks, and are spherical rather than cigar-shaped (Fig. 63A). Each one contains only a few ciliated sperms, which are about as large and elaborate as any sperms in the plant kingdom (Fig. 63B). The less numerous female sex organs are again

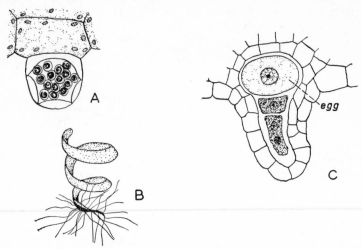

FIG. 63.—A, longitudinal section of male sex organ of fern. B, single sperm, enlarged. C, longitudinal section of slightly immature female sex organ of fern. The mature female sex organ will probably have its neck bent even more, and the cells now forming the core of the neck will have disintegrated, thus clearing a continuous passageway through which the sperm may enter.

constructed on the principle of a single large egg within the swollen base and a long sterile neck. In this case, however, the base is bedded back into the tissues of the gametophyte proper, and the neck, considerably shorter than that in mosses, is bent backward, away from the notch of the gametophyte (Fig. 63C).*

With the breaking of the wall of the male sex organ, the sperms find themselves in the thin film of water that is commonly present on

* In most ferns the mature gametophyte carries male sex organs near the pointed (posterior) end and female sex organs near the notch (anterior) end. This results from the following facts.

The very young gametophyte which has just emerged from the spore has the shape of a short filament (Fig. 61). Very soon, however, the growing-point

the soil surface. As is so common in both plants and animals, the sperms are influenced by a chemical exuded through this film of water from an egg in the neighborhood.† As usual, the first sperm to

---

establishes itself at the base of a notch, as in liverworts. Thereafter the growth of the gametophyte is based upon the activities of this growing-point, which keeps giving off (by cell divisions) new cells that enlarge to increase both the width and the length of the gametophyte. In this process the notch is continuously carried forward, and the heart-shaped contour of the gametophyte is maintained as the gametophyte enlarges.

When the gametophyte is still comparatively young (i.e., a small heart) it produces a few score of male sex organs. As the gametophyte continues to grow, the notch leaves these male sex organs behind. Later, when the gametophyte is approaching its full size, a half-dozen or so female sex organs are developed in the region which is at that time just behind the notch.

This difference in the time of production of the two types of sex organs is probably correlated with the nutritive capital of the gametophyte at the time of their production. Sperms, which are relatively small, may be produced on the basis of the nutritive capital of a young gametophyte. Eggs, which are large and packed with nourishment, can be produced only by a gametophyte which has had time to lay up a comparatively large nutritive capital.

One might be tempted to account for the backward bending of the necks of the female sex organs as an adaptation to facilitate the entrance of sperms released by the male sex organs of the same gametophyte. But self-fertilization is the exception rather than the rule in ferns. The male sex organs of a given gametophyte have usually discharged their sperms before the female sex organs of the same gametophyte are ripe to receive sperms. The eggs of the latter can be fertilized, therefore, only by the sperms of other, younger gametophytes in the neighborhood.

The backward bending of the neck is probably no more than the inevitable result of obvious mechanical forces. The neck of a young female sex organ protrudes straight down, and imbeds itself among the soil particles. Since the gametophyte is still elongating, and since a part of this elongation is the elongation of the cells around this particular sex organ, the sex organ is "dragged forward" and the imbedded tip of its still elongating neck is "left behind."

† These chemicals, which are known in a few cases to be organic acids, differ in different types of plants, the sperm of a given species being responsive to only one type of chemical (or perhaps to a limited number of types). If the sperms of a moss and of a fern are thoroughly mixed together in a small container of water, they may be separated by introducing at one point in the container the tip of a small pipette containing the chemical that is characteristic of either moss or fern eggs. This will attract only the sperms of one type and thus segregate them out of the mixture.

arrive fertilizes the egg, the outer membrane of which is thereby altered so as to exclude sperms that may arrive subsequently.

Just as the spore was the first cell of the gametophyte generation, so the zygote is the first cell of the sporophyte generation. As in Bryophytes, the fern zygote initiates at once the long series of cell divisions that is to culminate in the production of the mature sporophyte. A young, developing organism, more than any other, needs a supply of energy and materials. In the life of the fern sporophyte

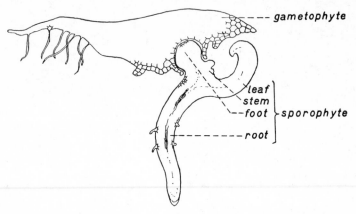

Fig. 64.—Longitudinal section through young fern sporophyte while it is still completely dependent on the gametophyte.

one may recognize three (somewhat overlapping) stages with respect to the sources of energy and materials.

In the beginning the very young sporophyte relies upon the cytoplasm of the zygote itself. This cytoplasm, which was the cytoplasm of the egg, is rich in materials which may be transformed into new structures or "burned" to yield energy. Even so, this supply cannot carry the developing organism far. If the young sporophyte is to succeed it must at once perfect arrangements for securing nutrition from some other source.

As among the ancestral Bryophytes, the thing that is emphasized in the first stages of growth is the foot. By pushing this foot well up into the overlying tissues the young sporophyte establishes a parasitic relationship with the gametophyte. It cannot truly be said, therefore, that the sporophyte of the fern is a completely independent plant, for there is always this temporary stage of dependency on the gametophyte. During this stage there is a rapid development of those sporophyte structures which will be needed to put the plant on an independent basis; for, again, the supply of nutrition provided by the tiny gametophyte is rapidly exhausted (Fig. 64).

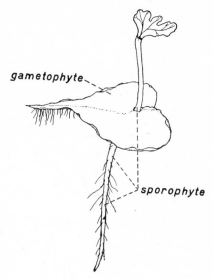

A primary root is pushed down into the soil, and a first leaf—absurdly small as compared with the leaves of the adult fern—coils around through the notch of the gametophyte and emerges into the sunlight (Fig. 65). The sporophyte is now "on its own." Food manufactured by the leaf supplies energy and materials for the development of the stem, which has been neglected up to this time. No longer of

FIG. 65.—Sketch of stage slightly later than that in Figure 64. The first leaf of the sporophyte has now found its way into the sunlight.

service, the foot withers away, as does the parent-gametophyte in most cases. Steadily the stem increases in size, the roots in both size and number. Larger and larger leaves are put out, and the sporophyte is well launched on its way toward the adult condition.

CHAPTER XII

HORSETAILS AND CLUB MOSSES

IN THOSE early days when the Pteridophytes dominated the earth's surface, the horsetails and club mosses were apparently represented by many species, some of which had tree-like bodies. Today these two groups have been pushed even farther into the background than have the ferns. Only a few species of each group remain; only in rare localities does one encounter them; and their sporophyte bodies, instead of being tree-like, are comparatively puny. One gathers the impression that horsetails and club mosses are archaic types, well along on the road toward extinction.

The horsetails (order Equisetales) are today represented by only the single genus *Equisetum*, containing about twenty-five species. All are characterized by the same unique sporophyte body, with several features which distinguish it sharply from the bodies of other vascular plants. The main stem is horizontal and subterranean, like that of most modern ferns. Unlike the fern, however, the horsetail sends up aerial branch stems of a striking type.

The fossil record tells us that many of the ancient horsetails developed sizeable leaves. In living forms, however, the leaves are mere scale-like vestiges, quite inadequate to supply food for the entire body. Instead, food manufacture is conducted in the aerial branches themselves, which carry chloroplasts in a zone slightly below the epidermis.* In these branches we encounter a conspicuous, jointed effect, roughly suggestive of the bamboo. This is due to a sharp differentiation of the branch into alternate "nodes" and "internodes." The capacity to produce lateral members is restricted to the very short nodes. Thus, a single "whorl" or "cycle" of tiny leaves is

---

* The transfer of food manufacture from leaves to stems is likewise encountered in several groups of the seed plants, notably the cacti. This maneuver markedly reduces the surface area of the plant in proportion to its volume. By thus reducing the amount of water lost through evaporation, it provides one method of adapting the plant to life in comparatively dry situations.

produced at each node, while the long internodes in between are entirely devoid of leaves. By pulling, one can rather easily separate these aerial branches into sections, the breaks occurring at the nodes (Fig. 66*A*).

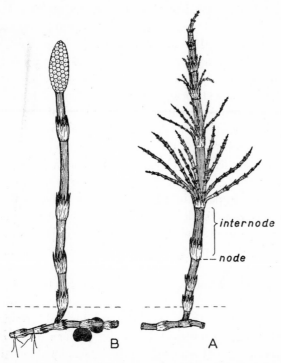

FIG. 66.—The sporophyte of *Equisetum*. *A*, portion of the underground main stem, giving rise to the aerial "vegetative" branch, which, in turn, is giving off numerous subordinate branches from its upper nodes. *B*, a "fertile" branch, surmounted by its strobilus.

In popular parlance the horsetails are often referred to as "scouring rushes." A handful of the aerial branches provides an abrasive which is quite effective for scouring pots and pans, for these branches, instead of being perfect cylinders, are fluted in contour, and their epidermal layers are impregnated with a deposit of silica.

In some species there is but one type of aerial branch, which takes care of both food manufacture and spore production. In others these two functions are allotted to separate types of aerial branches, which

are produced at different times in the season. In such cases the subterranean main stem will put up only the green, vegetative branches
during the summer. In these the surface for food manufacture is increased by the production of subordinate branches (smaller replicas
of the primary branches), which come off at the nodes in the "axils"
of the diminutive leaves (i.e., in the angles between the leaves and the
main axis). The vegetative branches die away in the fall, but much
of the food that they have produced has been stored in the underground stem. On the basis of this food supply the stem throws up
fertile branches the next spring. These are not self-supporting, being
unbranched and provided with relatively little chlorophyll. At the
tip of each fertile branch is a "strobilus" (cone), a new type of reproductive structure, in which the spores are produced (Fig. 66B).

A spore-bearing leaf is technically known as a "sporophyll." A
strobilus is a rather compactly organized group of sporophylls. Some
form of strobilus appears in all of the plants to be discussed here-
after. In some, as we shall see, the leaf-like character of the sporophylls is still apparent. In others, the sporophylls have become so
highly specialized for spore production as to bear little resemblance
to an ordinary leaf.

This last is conspicuously true of the sporophylls of *Equisetum*.*
Here the sporophylls consist of simple cylindrical stalks which come
off at right angles from the axis of the strobilus and terminate in expanded shield-like structures. The outer surface of an immature
strobilus is composed of hundreds of these hexagonal "shields" with
their edges fitted together perfectly (Fig. 66B). From the inner surface of each shield there extend five to ten finger-shaped "sporangia"
(spore-cases), arranged in a ring around the stalk of the sporophyll
(Fig. 67A). At the stage of maturity the shields pull apart at their
edges, slits appear in the sporangia, and the spores are thus released
into the air.

The spores of the horsetail were objects of fascination for the early
microscopists. In addition to the surrounding wall present in most

---

* In fact, study of a series of fossil forms suggests that the true sporophyll has
been completely suppressed in *Equisetum*, and that the so-called sporophyll is
in reality a modification of another structure ("sporangiophore") that was
originally associated with the sporophyll.

spores, those of the horsetail possess an outer wall which acts in an amazing way. At the time of spore-shedding this outer wall cracks in a very regular pattern into two ribbon-like strips, arranged spirally around the spore. Each strip remains attached, in its central region, to one pole of the spore, but elsewhere becomes loosened from the spore surface. At shedding, therefore, each spore, like a maypole, carries these four radiating streamers (Fig. 67*B* and *C*). The streamers are "hygroscopic," coiling and uncoiling in response to slight changes in the humidity of the atmosphere. The net result of it all is that *Equisetum* spores become tangled with each other and fall in

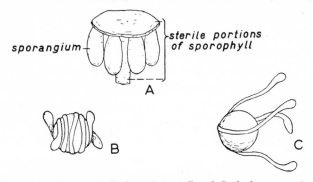

FIG. 67.—*A*, single sporophyll of *Equisetum. B* and *C*, single spores, showing the peculiar attached strips coiled and uncoiled.

clumps rather than singly. Perhaps this peculiar feature has adaptive value. Certainly it increases the likelihood that several gametophytes will grow in close proximity; and this, in turn, favors fertilization, for the gametophytes of this form are differentiated into male and female individuals.

Though all the spores are of the same size, half of them produce male gametophytes and the other half produce female gametophytes. Both gametophytes are flat, green bodies, roughly suggesting those of the fern. Apparently, however, these gametophytes follow a very disorderly program of development, branching wildly and having at maturity a highly irregular contour. Male and female gametophytes are alike in general appearance, but the female is several times as large as the male. This feature doubtless bears a functional relation to the higher nutritive content of the female gamete, and to the fact that the young sporophyte develops for a time at the expense of the female gametophyte.

The remainder of the life-cycle of *Equisetum*, though peculiar in some of its details, is in the main like that of the fern. Large, ciliated sperms, released from the male sex organs of the male gametophyte, swim through films of water to reach the female sex organs of the female gametophyte. The zygote yields a sporophyte, which pushes a small foot back into the tissues of the female gametophyte, thereby extracting the nutrition needed to carry it through its early stages of development. After a root has established adequate soil relations and green branches have been pushed up into the sunlight, the sporophyte embarks upon an independent existence.

The living club mosses (order Lycopodiales) are remnants of what was probably the oldest of Pteridophyte groups. In theory, one can visualize a gradual transformation of the stalk-like green sporophyte which occurs in the *Anthoceros* group of liverworts into the rather simple leafy stems which are present in some of the living Lycopodiales. No other derivations of Pteridophytes from Bryophyte ancestors can be so readily comprehended. In the fossil record, too, we find the club mosses to be among the earliest Pteridophyte types. Like the horsetail assemblage, this group also was once responsible for forests of tree-like forms whose transformed bodies contributed to the coal seams of today. Though the Lycopodiales, too, have suffered a decline, they are today somewhat better represented than the horsetails, having four living genera and some hundreds of species.

The two most prominent genera are *Lycopodium* and *Selaginella*. Since the former shows primitive characteristics and the latter has become distinctly more modernized, the life-cycles of the two will be treated separately.

Lycopodium, which contains larger sporophyte bodies than *Selaginella*, is often encountered in the woods of the northern United States. The main stems commonly sprawl prostrate along the surface of the ground, sending up erect branches to the height of a foot or more. As all the stem parts are copiously coated with small, sharp-pointed leaves, *Lycopodium* is often called the "ground pine."

Some species of *Lycopodium*, apparently more primitive than the rest, lack definitely organized strobili. Instead, spore production occurs at many points along the erect branches. Though all the leaves are superficially alike, some of them are actually the sporophylls, for they carry solitary, kidney-shaped sporangia on their upper surfaces, near the points where they attach to the stem. In the majority of species, however, all leaves on the main stem and the lower regions of the erect branches are strictly vegetative, and the sporophylls are closely organized into strobili at the upper tips of the branches. In all species the leaf-like character of the sporophyll is quite plain (Fig. 68).

All spores of *Lycopodium* are of the same type. In size the spores are among the most constant of biological objects, and it is for this reason that the early microscopists often mounted a few *Lycopodium* spores on the same slide with objects which they wished to measure.

Unlike the horsetail, *Lycopodium* produces only the one type of gametophyte. Very few collections of botanical material contain *Lycopodium* gametophytes, for they are extremely difficult to find. In many species the gametophyte, after it has been produced from the spore, works its way down into the ground. Thereafter it maintains a subterranean, saprophytic existence—a tiny, irregular, tuberous affair. In other species the gametophyte, though subterranean in the main, is near enough to the surface to push some of its tissue into the air and conduct photosynthesis. In a few species the gametophyte is entirely aerial.

Once again eggs are fertilized by swimming sperms. Once again the young sporophyte that comes from the zygote is parasitic for a short time upon the tissues of the gametophyte.

The genus *Selaginella*, very rare in this part of the world, is fairly common in the tropics, where it is represented by four hundred species or more. These are commonly called the "little club mosses," being consistently smaller and more delicate in texture than the *Lycopodium* forms. In general appearance, however, the sporophyte bodies are much like those of *Lycopodium*. At maturity there appear at the tips of the branches rather compact strobili composed of distinctly leaf-like sporophylls, each with a single sporangium on its upper surface, as in *Lycopodium*.

At this point, however, the resemblance ceases, for it is in *Selaginella* that we encounter for the first time the phenomenon of "heterospory." Heterospory, or the production of two different types of spores, was a feature of the life-cycle that meant a great deal to the evolution of the plant kingdom. Introduced some hundreds of millions of years ago among a few of the more progressive members

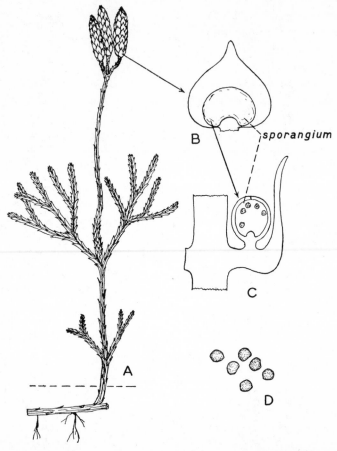

FIG. 68.—*Lycopodium.* *A*, the sporophyte body, bearing a cluster of strobili at the upper end. *B*, upper view of single sporophyll. *C*, longitudinal section to show side view of single sporophyll and its relation to the axis of the strobilus. *D*, spores.

of the then existing Pteridophytes, it opened the way for the subsequent introduction of the seed, that culminating reproductive achievement of the entire plant kingdom that appears only in the members of the highest group. Today heterospory is rare among the surviving Pteridophytes, most of which retain the more primitive "homospory," as we have seen. The heterospory of *Selaginella*, however, is so beautifully diagrammatic that this form is commonly used to give the student of botany a better understanding of the transition between Pteridophytes and seed plants.

The two types of spores in *Selaginella* are referred to as "megaspores" (large spores) and "microspores" (small spores). A megaspore has almost ten times the diameter (and a thousand times the volume) of a microspore. Both are produced in the same strobilus, but within different sporangia and on different sporophylls. Accordingly we say that the strobilus is made up of "megasporophylls" and "microsporophylls." Actually there is no difference whatsoever in the sterile or "leaf" portion of the two sporophylls, but the one carries on its upper surface a single "megasporangium," the other a single "microsporangium." The cream-colored megasporangium carries only four gigantic megaspores, far larger than any spores hitherto encountered. The reddish microsporangium carries a few thousand of the much smaller microspores. Both types are produced in the same strobilus, the megasporophylls usually (but by no means always) being carried lower in the strobilus and the microsporophylls nearer to the tip (Fig. 69).

A further *Selaginella* feature that is prophetic of the condition found in seed plants lies in the fact that the gametophytes are retained within (or practically within) the walls of the spores that produce them. Since this prevents exposure to sunlight, the gametophytes are necessarily dependent, their only supply of food being that which was stored within the spores. Thus the gametophyte, the dominant generation among Bryophytes, has dwindled in magnitude until at last it has lost its independence completely and become a rather inconspicuous part of the life-cycle.

While still retained within the microsporangium, a microspore produces a tiny male gametophyte, which consists of no more than a single vegetative cell plus a single, small but many-celled male sex organ, the entire gametophyte being completely contained within the microspore wall. While still retained within the megasporangium, a megaspore produces a female gametophyte, small by comparison with the gametophytes that we have seen in other forms but vastly larger than the male gametophyte. The female gametophyte is not quite contained within the megaspore wall, but protrudes slightly at one point where the spore wall becomes ruptured. The bulk of this gametophyte is made up of many vegetative cells, well

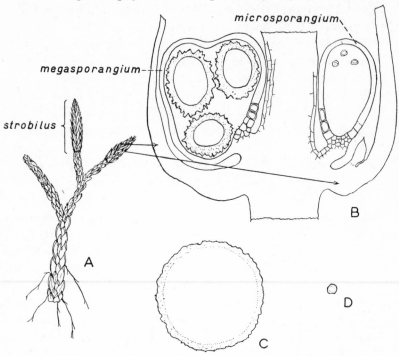

FIG. 69.—*Selaginella. A*, the sporophyte body. *B*, longitudinal section of portion of strobilus, including one megasporophyll and one microsporophyll. *C* and *D*, megaspore and microspore, drawn to the same scale.

stocked with a food supply for the benefit of the embryo sporophyte that is to come. The protruding portion of the gametophyte carries a few female sex organs (Fig. 70).

Once again fertilization depends upon the swimming of a ciliated sperm through a film of water, but this no longer occurs in the customary setting of the soil surface. The sporophylls, which up to this time had been held rather tightly in the compact strobilus, unfold somewhat, like the petals of an opening flower, so that the sporangia on their upper surfaces now become well exposed. The megasporangium walls crack open, but still retain the megaspores, like eggs in a nest. Microsporangium walls rupture and release showers of microspores, some of which lodge among the megaspores of sporophylls which were lower in the same strobilus. It is at this stage that the microspore wall ruptures, as does also the wall of the male sex organ. The sperms, now released, swim through a film of water to the exposed female sex organs nearby. Thus, within the strobilus itself, there appear many embryo sporophytes, each growing parasitically at the expense of the female gametophyte which is still largely buried

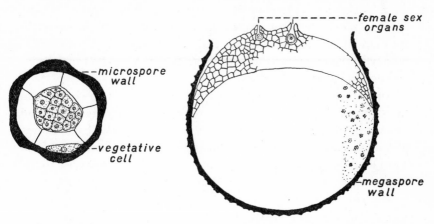

FIG. 70.—The gametophytes of *Selaginella*. On the left, the tiny male gametophyte with its single sex organ, consisting of a surrounding layer of sterile cells and a packed mass of spermatogenous cells within. On the right, the much larger female gametophyte.

within the old megaspore wall. A little later, whether megaspores fall singly from the withering strobilus, or whether the strobilus as a whole drops to the ground, some of the young sporophytes will be able to get primary roots into the soil, green leaves into the air, and thus become established for an independent existence.

In résumé of the Pteridophytes, we should emphasize the rise of the independent sporophyte generation and the roughly corresponding decline of the gametophyte. This independent sporophyte brings with it the body organization into roots, stems, and leaves, together with the new "vascular" or conducting tissue. We should also emphasize heterospory, attained by only a few of the higher Pteridophytes, and a necessary antecedent to the formation of the seed, which characterizes the next and last great group.

CHAPTER XIII

THE MODERN PLANT BODY

THE last of those four great groups which compose the plant kingdom is that of the "Spermatophytes," or seed plants. Judged by most standards this group is today more successful than any of its predecessors, and the success may be attributed largely to the high reproductive efficiency which results from the introduction of the seed.

Among seed plants the botanist recognizes the two co-ordinate divisions which we may describe roughly as "cone-bearers" and "flower-bearers."* The fossil record tells us that the cone-bearers are much the more primitive. Apparently these were derived from Pteridophyte ancestors long ago, and, in their turn, gave rise to the flowering plants at a much later date.

Most of the cone-bearers are trees and most of them are evergreens, bearing their leaves all year round instead of shedding them every winter.† The group contains such familiar forms as pine, spruce, fir, hemlock, and sequoia. These are the giants of the plant kingdom,‡ and constitute the larger part of our timber supply.

Success, however, depends not upon size alone. Ages ago the gigantic dinosaurs of the animal kingdom died out completely and left the earth to the smaller but more adaptable mammals and birds.

---

* A more accurate distinction between these two groups is described at the beginning of chap. xvii.

† The evergreen is continuously shedding a few of its older leaves, as is attested by the carpet of dead needles that we find in a pine forest. At the same time new leaves are continuously developing, so that there is never a time that the stems are completely denuded of their foliage.

‡ Many of the cone-bearers are trees of up to 200 feet in height. Biggest of all is the giant sequoia, which has been known to reach 400 feet. By comparison most flowering plants are puny affairs, though quite a few are trees, and some fairly large ones. One flowering plant, the eucalyptus of Australia, rivals the giant sequoia in size.

Perhaps a similar fate awaits our cone-bearing trees. For a long time in the past they dominated much of the earth's surface, but more recently have been giving away before the flowering plants, which are in general more rapid in growth, more efficient in reproduction, and more adaptable to a variety of living conditions. Today there remain only about 500 species of cone-bearers in existence, as against about 130,000 species of flowering plants.

The flowering plants may be characterized by a number of superlatives:

1. That they are the most advanced and most complex goes without saying, since they constitute the culmination of the evolutionary sequence that has been sketched.

2. That they are the most recent is not only inferred from their evolutionary position but demonstrated by the fact that they appear last in the geologic record.

3. That they are most diversified is to be expected of such an up-to-date assemblage. This is strikingly revealed by the variety of plant forms seen in the common natural panorama, which is likely to display to the untrained eye no other plants than flowering plants.

4. That they are economically most important is not only because they are the basis of our food supply but also because of the drugs, fibers, paper, and other textiles that they provide. In addition, they make a large contribution to man's wood supply.

The sporophyte body of all Spermatophytes maintains the same fundamental plan that was developed by the Pteridophytes. Roots, stem, and leaves, with woody conducting tissue running throughout, are still the essentials of body organization.* The body of the cone-bearer is usually a tree, sometimes a shrub. The body of the flowering plant may be cast in any one of a great many patterns, for the flowering plants live under an extremely wide range of environmental conditions and show corresponding adaptations.

Most flowering plants can be easily recognized as conforming to one of three general forms: "trees," in which the stem is large and erect; "shrubs," in which the stem is smaller and bushy; and

---

* Pteridophytes and Spermatophytes are together often referred to as our "vascular plants."

"herbs," in which the stem is tender, with woody tissue much reduced. Some stems, however, sprawl horizontally over the soil surface, and some remain buried beneath the surface, sending into the
air nothing but the leaves (and flowers). Other stems assume a
climbing or vinelike habit, while some even perch upon other plants
and maintain no soil connections of their own. A tropical forest presents a dense, tangled array of all these types, and is a striking illustration of how plants may insinuate themselves into every last little
niche that will somehow provide them with the materials and energy
which they require.

Quite a number of cone-bearers and some flowering plants can
withstand the rigorous climate of very high altitudes and latitudes,
though of course there are many more which can live successfully
only under tropical conditions. It is the presence of available water
that limits the distribution of plants more than does any other single
factor. The barrenness of deserts is usually attributable to this factor alone, as is evidenced by the rich plant growth that is usually
possible when irrigation is introduced. Some flowering plants maintain themselves successfully in near-desert conditions. These are
characterized by certain rather obvious adaptive features: highly
developed waterproofing layers; small, thick leaves, or none at all,
photosynthesis being relegated to the stems; very long roots; large
water reservoirs within the tissues. At the other extreme are a few
flowering plants that revert to life in a water medium.*

With all this variation in form, however, the many flowering
plants exhibit only the one fundamental body plan. Always there is
the organization into roots, stems, and leaves. Always there is the
woody conducting system running throughout the body.

Like all other parts of a living body, wood† is a product of living
cells. But the wood itself, at the time that it is serving the organism,
does not consist of living cells. In the production of wood, certain
cells are considerably enlarged and their walls impregnated with a

---

* These forms show adaptive features, but retain the major characteristics
of their more recent ancestors, rather than reverting to the body plan of the
algae. In the plant and animal world we have little or no evidence of evolution
simply turning backward along the progressive paths it has traversed.

† Or, more technically, "xylem."

substance* which makes for rigidity, with a measure of elasticity. When these cells are fully developed they die, and their protoplasm disintegrates. It is in this dead condition that the wood cell serves the plant, for its skeleton provides a hollow tube through which water and soil salts can be transported. As a rule many of these wood cells are "welded" together in a series, and the end-walls disintegrated to form a long, continuous pipe or "vessel." In the woody

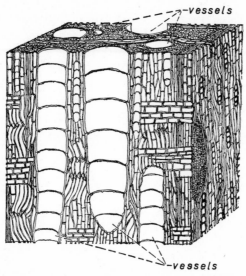

FIG. 71.—Three-dimensional enlargement of a piece of wood, showing the occasional large vessel embedded in a tissue of smaller caliber cells.

portion of the stem of the plant one finds many of these large conducting vessels imbedded in a mass of smaller cells.† This is the structure that would be revealed by microscopic examination of any piece of wood, whether that wood had been recently a part of a living plant, or whether it had been applied to man's uses thousands of years ago (Fig. 71).

---

* This substance is known as "lignin."

† Though some of these smaller cells probably also engage in a limited amount of water conduction, their more conspicuous biological value is in providing the plant body with a degree of strength and rigidity that it would otherwise lack.

To serve the plant efficiently as a transporter of water, this wood must run not merely through the stem but through roots and leaves as well. Water absorbed near the tips of the roots passes into the lowermost ends of the woody tubes. From that point it moves upward through a continuous series of these natural pipes, through the

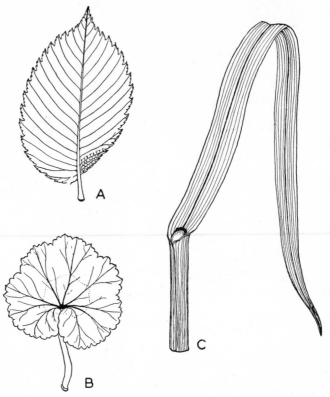

Fig. 72.—Leaf venation. *A* (the elm) and *B* (the common geranium) are "net-veined," while *C* (corn) is "parallel-veined."

length of the root, through the main stem itself, and through all the branches. Just as the main, "cable"-like mass of wood cells in the stem "breaks up" into smaller bundles that lead out into the branches, so also the bundles within the branches yield still smaller bundles that run out into the leaves. This process of subdivision continues within the leaf itself, and there we can usually follow it by superficial inspection, for the bundles of woody conducting tissue are represented by the visible "veins" of the leaf. One sees how the main

vein breaks up into progressive- ly smaller elements, which run through all parts of the leaf and thus provide for an efficient de- livery of water to every living cell (Fig. 72).* To get any ade- quate picture of the whole plant as a working unit, therefore, one must think not merely of wood as a tissue but also of its organ- ization into the elaborate and much ramified "vascular sys- tem."

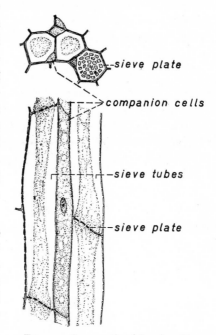

The vascular system, however, includes a second component that is as necessary as the wood. Wood provides merely for the conduction of water and soil salts from the roots to the other parts of the plant. Provision must also be made for a conduc- tion of the food that is manu- factured in the leaves to the many other parts of the plant that are unable to manufacture food for themselves.

FIG. 73.—Cross-section (above) and lon- gitudinal section (below) of a bit of phloem. It is the so-called "sieve tubes" that con- duct the food, passing it along from cell to cell through the "sieve plates." The func- tion of the associated "companion cells" is not well understood.

The transportation of food is carried on by another tissue that is known as "phloem." Like the wood cells, these phloem cells are

---

* With respect to their "venation" (arrangement of veins in the leaf), the two great subdivisions of flowering plants differ. Plants of the one subdivision (the dicotyledons, viz., chap. xx) have "net-veined" leaves, in which larger veins break up into progressively smaller branches, as described above. Those of the other subdivision (monocotyledons), such as the grasses, have "parallel- veined" leaves, in which all veins are of about the same magnitude and run almost parallel with each other from the base of the leaf to its tip.

elongated, pipe-like, but they do not have the thick walls of the wood. The end-walls between successive phloem cells are not com-pletely dissolved away, as in the wood, but, sieve-like, contain a number of small apertures through which the food apparently passes. Another difference lies in the fact that the phloem cells are not dead, but actually contain a small amount of living protoplasm (Fig. 73). Even so, these phloem cells pro-vide channels which are adequate for the transport of food. The upward move-ment of water and soil salts through the wood is much more rapid than the downward movement of food through the phloem.

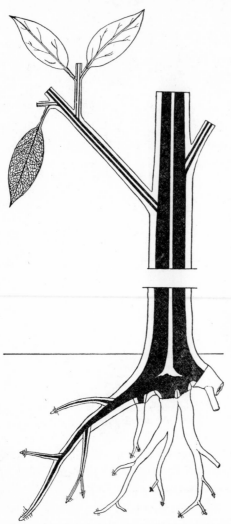

FIG. 74.—Idealized diagram of the modern plant body, to show how the vascular system (darkened) traverses roots, main stem, stem branches, and leaves.

Like the wood, the phloem is organized into a connected system which extends throughout the entire plant. The arrange-ment of the system is easily described, for it corre-sponds exactly with that of the wood. Wherever in root, stem, or leaf there is a bundle of wood cells, right beside it is a (usually smaller) bundle of phloem cells. Structurally, in fact, the two components are united so compactly that botanists common-ly speak of a "vascular bundle" as a structural unit that contains both wood and phloem components. Correspondingly, when the botanist says, "vascular system," he is referring to the entire conducting system of the plant, with its parallel elements of wood and phloem (Fig. 74).

The development of the complex and often tremendous body of the seed plant, with its various parts so beautifully adapted to the environment in which each exists, yet all so neatly co-ordinated into an efficient working unit, is a phenomenon which has always impressed the biologist deeply and challenged his understanding. This entire sporophyte body has been derived by a series of cell divisions from a single-celled zygote. As we shall see later, the sporophyte is, at an early stage, a tiny and apparently quite simple affair that is contained within the seed. Emerging from the seed, it establishes its first little roots in the ground and pushes its diminutive stem and early leaves up into the air and sunlight. From this stage on, the increase in size is to be accounted for in the main by the continuous division of cells in certain critically placed growing regions of root, stem, and leaf, followed by the growth and gradual transformation or "differentiation" of these cells into the adult condition of the various tissues of which they are to form a part.

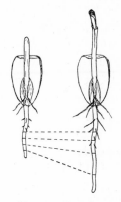

FIG. 75.—Experimental demonstration that the elongation of the root occurs near its tip. Evenly spaced markings were made on the root of the young corn seedling at the left. In the ensuing twenty-four hours the elongation of the root separated the markings in the manner shown on the right.

In most flowering plants the sporophyte body displays a conspicuous primary axis with stem tip at the upper end and root tip at the lower. Growth in length is to be accounted for by active cell division in tiny zones right at the very tip regions.

It is only at the very tip of the stem, therefore, that one finds cells actively dividing; just behind this a zone in which the young cells are enlarging; and still farther back a region in which the cells are beginning to be differentiated into the various tissues. In short, the stem adds to its length by building on more cells at the top and articulating them with the tissues that are already present.

In the root one encounters corresponding zones: a tiny zone of active cell division at the tip, a zone of elongation just behind, and a zone of differentiation next. This growing region of the root exhibits one conspicuous feature, however, which is not present in the stem. The stem in its elongation is forever pushing the tender young tip region up through a medium of air, but the root must thrust its tip through the soil. Inevitably this would bring an abrasive action by the soil particles—an action which would destroy the only portion of the root that was capable of elongation—were it not for an adaptive device which is peculiar to the root. The little zone of dividing cells at the root tip, while continuously laying down behind itself those cells which are to be incorporated into the root proper, at the same time keeps laying down a few cells which are thrust out in front to form the so-called "root cap." The root cap acts as an effective buffer or helmet for the root tip proper, but is itself inevitably worn away as it is thrust through the soil, so that its tissues must be renewed continuously by the root tip. In a longitudinal section of a root tip, therefore, one finds: root cap at the very bottom; next, the very small zone of tiny, actively dividing cells; next, a zone where the cells are enlarging to attain their adult size; and then a zone where the enlarged cells are being matured and differentiated to assume the adult characteristics of the various tissues (Figs. 75, 76).

Save in the region of the root cap, the most superficial tissue is the epidermis. As one might expect, this lacks the thick walls and waterproofing material that characterize the epidermis of aerial parts, for the total supply of water and mineral salts that the ordinary plant will ever get is what enters through the epidermis of the root. Absorption is the primary function of the root, and the absorbing surface is greatly increased by the production of "root hairs." Throughout a zone of perhaps an inch in length,* starting a short distance back of the root tip, a good many of the epidermal cells

---

* The length of this zone depends upon the magnitude of the root as well as upon other circumstances.

push out these long protrusions, or root hairs, which establish an intimate contact with the soil particles. It is this short root-hair

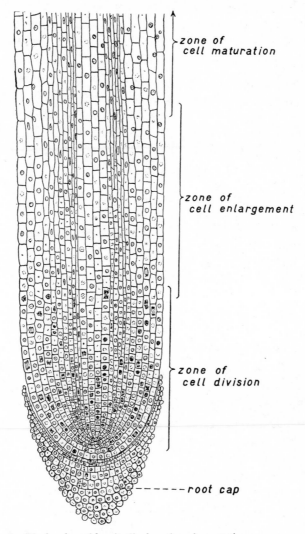

zone of
cell maturation

zone of
cell enlargement

zone of
cell division

- - - - - root cap

FIG. 76.—Much enlarged longitudinal section of a root-tip

zone that accounts for practically all of the absorption that occurs (Fig. 77).

It should be noted that the root hairs, like the cells of the root cap, are ephemeral rather than permanent. The life of an individual root hair is only a day or so. New root hairs are continuously being produced at the "younger" edge of the root-hair zone; old ones are continuously withering away at the "older" edge of the zone; and so, as the root pushes on through the soil, a zone of active root hairs is steadily maintained a short distance behind the advancing tip. The older regions of the root are completely devoid of root hairs, and, for other reasons as well, are quite ineffective in absorbing water and salts.

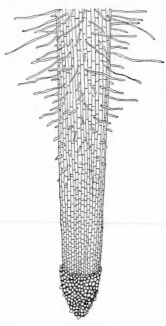

FIG. 77.—Enlarged surface view of root-tip, including root-hair zone.

Branch stems and branch roots are small replicas of the main branches and roots. Each is provided with its own growing tip; each may, in turn, give rise to secondary branches. The details as to the origin of the branches and the account of how their vascular bundles are made continuous with those of the main stem and root are matters beyond the scope of the present book. One point, however, should be noted, for it provides a practical distinction between stem and root.* Branch roots may come off at practically any point along the main root (Fig. 78A). Branch stems, however, will be produced only at certain points on the primary stem. The stem is divided along its length into alternate "nodes" and "internodes," the nodes

---

* If all roots grew underground and all stems in the air, even the untrained person would find little difficulty in making the distinction. Since some roots grow in air, however, and quite a number of stems grow underground, one can identify these organs only by means of their distinctive structural features.

being exceedingly short zones (or merely tiny patches of tissue) and the internodes being the relatively long zones which extend

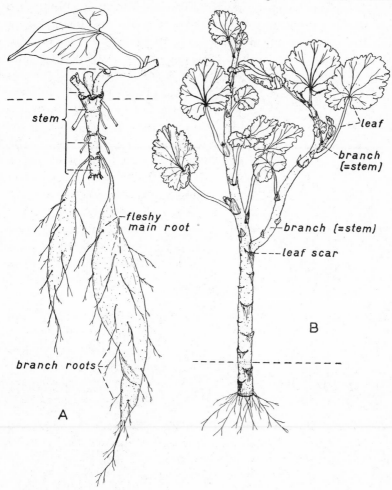

stem

—leaf

—branch
(=stem)

—fleshy
main root

—branch (=stem)

----leaf scar

branch roots

A

B

FIG. 78.—*A*, lower portions of a young sweet potato plant, showing the random production of branch roots from the fleshy main root. *B*, the common geranium plant, showing how branch stems are produced only at the nodes, where they arise in the angle between the leaf and the main stem.

between successive nodes. The power to produce lateral members is restricted to the nodes. Branch stems, leaves, and flowers, therefore, will arise only at these points in the stem and not elsewhere (Fig. 78*B*).*

When it comes to the leaf, increase in size cannot be accounted for by the activity of a small group of cells at the tip. In the typical leaf, cell division and enlargement appear to occur at the same rate throughout all of the tissues. This fact is rather neatly demonstrated by the simple experiment recorded in Figure 79.†

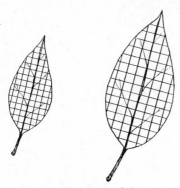

FIG. 79.—Experimental demonstration that enlargement of a typical leaf is not restricted to any particular zone. On the left, a young leaf to which "cross-hatching" has been applied with India ink. At the right, the condition of the same leaf a few days later.

In connection with their elongation, stems and roots exhibit a remarkable series of adaptive responses. As a rule the body of the living plant is surprisingly well adjusted to the environmental conditions in which the species ordinarily lives. This is true not only of the body as a whole but of the various parts of the body, regarded separately. In nature, therefore, we usually find the plant so oriented as to derive a maximum of good and a minimum of harm from the environment; roots burrowing into the soil "in quest of" water and soil salts, stems elevating the leaves and exposing them to the sunlight. The attainment of this adaptive orientation in the first place, and its subsequent modification in re-

---

* What we call the "sweet potato" is indeed a root, devoid of nodes and producing a few diminutive branch roots from various points along its length. What we call the "Irish potato" or "white potato" is a stem. Its character is revealed by the so-called "eyes," which are actually the nodes. Under certain conditions branches develop from these nodes, but not from any other point on the potato.

† The description given above applies to the net-veined leaf of the dicotyledon (viz., footnote on p. 139). In the parallel-veined leaf of the monocotyledon, cell division and elongation is largely restricted to a zone near the base. For this reason grass blades, clipped short by the lawnmower, will continue to elongate if their basal zones are intact.

sponse to changes in environmental conditions are the results of a group of plant responses that the botanist refers to as "tropisms."

If a group of young "seedlings" (sporophytes which have recently emerged from the seeds) is grown in a pot and exposed to light from above, all the young stems will push straight upward in their growth. Place them now in such a position that light reaches them only from the right side, and it will be noticed within a day or less* that all the young stem tips have turned and are now growing toward the new source of light. Clearly this is not an active muscular movement, such as characterizes the higher animals, but is a slower adjustment that is dependent on "differential growth." It can be demonstrated that the stem tip, exposed to light from the right only, elongates more rapidly *on the side away from the light* (i.e., on the left) than on the side toward the light. This differential growth results in a gradual reorientation of the stem tip, and ceases as soon as the stem tip is so oriented as to be receiving as much light on the right as on the left. This response of the plant is a tropism.† Since the stimulus in this case is light, this particular type of tropism is called "phototropism." Since stems in general grow toward rather than away from the light we say that they are "positively phototropic."

At the same time most stems are "negatively geotropic." "Geotropism" is a differential growth response to the stimulus of gravity, and it is not difficult to demonstrate that most stems will grow away from the "pull" of gravity (or away from any pull—i.e., centrifugal force—that is equivalent to that of gravity). Positive phototropism and negative geotropism co-operate in attaining and maintaining for the stem that orientation which is most advantageous for the stem and the leaves that it carries.

Roots are "positively geotropic." Pointed at first in a different direction, the root, through differential growth in the root-top re-

---

* The time required depends in the main upon the rate of growth that characterizes the species and the particular stage in the life of the individual at which the experiment is performed.

† The zoölogist uses "tropism" to refer to a type of response among simple animals for which the underlying mechanism is quite different but the results much the same.

gion, will redirect itself and grow downward, whether or not this maneuver succeeds in plunging it into the soil. Roots are also "positively hydrotropic"; i.e., they will grow toward a source of water, be that in the soil or elsewhere. Clearly these two tropisms tend to bring the root into the most effective relationship with the available environment (Fig. 80).*

Mere growth in length, however, will not account for all of the size increase that occurs in plants. The majority of stems and roots increase in diameter also. This, too, is due to cell division in a restricted region, but an understanding of the process requires some knowledge of the cross-section of the stem.

If we cut a horizontal slice through a young stem and examine it under the microscope, we are at once impressed with the concentric, target-like arrangement of the tissues. At the outside is the epidermis, a single, compact layer of cells, endowed with heavy waterproofed walls to cut down the loss of water from evaporation. Next inside comes the "cortex," a zone consisting of several layers of relatively unspecialized cells in which a certain amount of food storage occurs. In the young stem some of the outer cells of the cortex con-

---

* As the result of positive geotropism roots grow toward moist regions of the soil. It is moist but aerated soil, rather than the complete water medium, that represents an "optimum" (most advantageous) condition for the root of the ordinary land plant. The specialized roots of water plants thrive when submerged in water, but those of land plants are much less likely to do well under conditions of complete submergence. Roots, like other tissues, need oxygen, and in the ordinary land plant much of this oxygen is apparently taken in from the air which is present between the soil particles. Complete submergence, by eliminating the air and reducing the oxygen supply, may thus "suffocate" such roots as are not adjusted or adjustable to a medium of water.

Growth-responses to light and to gravity do not always direct the responding organ toward or away from the stimulus. Very commonly branch stems appear to be "transversely geotropic," extending in a more or less horizontal direction. Leaf blades are usually "transversely phototropic," presenting their flat surfaces toward the source of the maximum light. In this last case the most important factor is usually differential growth on the part of the "petiole," or small stalk which attaches the blade of the leaf to the stem.

Growth-responses to other stimuli are also exhibited by various plants. A rather remarkable case is that of "thigmotropism," the growth-response to contact or friction, which causes twining plants to coil around their supports.

tain chloroplasts and conduct food manufacture for a time.* Next within comes a zone which includes the phloem, together with certain other specialized tissues. Within the phloem is the wood, but between the two lies a single layer of cells known as the "cambium." It is the cambium, as we shall see, that is largely responsible for the stem's increase in diameter. Within the wood, at the very center, is a core of loosely arranged, unspecialized cells known as the "pith" (Fig. 81).

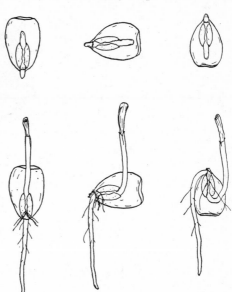

In the young stem all of the aforementioned tissues have been derived from the region of cell division at the stem tip, but the bulk of tissue that one finds in the cross-section of an old stem has been derived secondarily from the cambium layer. If one were able somehow to dissolve away all other tissues, the cambium of the entire stem would be seen to have the configuration of a very thin-walled stove-pipe. Actually the cambium is in close contact with a zone of wood on the inside and a zone of phloem on the outside. Through the life of the plant the cambium cells retain the capacity to divide and form daughter-cells.

FIG. 80.—Experimental demonstration of the negative geotropism of the stem and the positive geotropism of the root. Sprouting corn seedlings were oriented in the manner shown in the row above. About forty-eight hours later they had reached the condition represented in the row below.

When a cambium cell divides, one of the resulting daughter-cells carries on the characteristic cambium capacity, while the other

---

* At this same stage the epidermis is punctured by a few breathing-pores— a necessity for food manufacture.

gradually becomes differentiated and adds itself to one of the adjoining tissues. At a given division it may be that the *outer* of the two daughter-cells continues to be cambium, while the inner of the two gradually becomes specialized into a wood cell and adds itself to the mass of wood cells that is lying just inside. A little later the cell

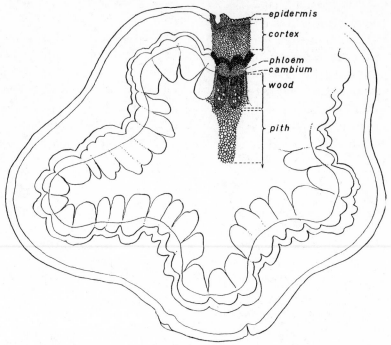

FIG. 81.—Cross-section of young oak stem (less than one year old)

which is now the cambium will divide, and in this case the *inner* of the two daughter-cells may continue to be cambium, while the outer of the two adds itself to the adjoining phloem. In this way the cambium, "working alternately with both hands," continues to increase the amount of wood and phloem as the stem grows older.* In

---

* Though the account given above may fairly depict the principle of cambium activity, it is not quite accurate in detail. Actually there may at times be several new wood cells formed in succession before a phloem cell is formed. The net result is that in the course of a growing season the total number of new wood cells added is markedly greater than the total number of new phloem cells.

this action all of the cells of the cambium cylinder keep pace with one another (under ordinary circumstances), so that the stem as a whole, though steadily increasing in diameter, maintains its cylindrical form.

Inevitably the addition of new wood within forces outward not only the cambium itself but all tissues which lie outside the cambium; while the addition of new phloem is a second effect which forces outward the old phloem, the cortex, and the epidermis. A time is soon reached when the epidermis, stretched beyond the limit of its elasticity, is ruptured. The resulting exposure of moist cells to the drying influence of the atmosphere might work a great deal of harm to the plant were it not that the exposed cells of the cortex meet the situation with an adaptive response. A single layer of cells, at or near the outer edge of the cortex, now becomes the "cork cambium." "Working with both hands," as did the main cambium, it proceeds to lay down several layers of "cork" on the outside and adds new cortex cells on the inside. Cork is a tissue which consists of tabular layers of cells, each endowed with thick walls that are impregnated with a waterproofing material.* As a protection against loss of water it is probably even more effective than the original epidermis. The first cork layers may themselves be ruptured by the increasing diameter of the stem, in which case newer and younger zones of cork will form beneath them. The roughness of the bark of many trees is the result of this repeated rupturing of the surface and the formation of new cork within.

In examining a cross-section of an old stem, therefore, one would look in vain for the epidermis. In its place there is a zone of cork, of which the thickness and the roughness will vary with the species of tree examined. Next beneath there may remain some unspecialized cortex, next the phloem, and next the single cambium layer, which has continued until death at its work of producing new phloem and wood. Though wood may constitute only a relatively small area in the young stem, the cross-section of the old stem presents a very different picture. Now by far the greater bulk of the cross-section area consists of wood, beside which the pith at the center and the

---

* This material is known as "suberin."

tissues outside the cambium appear to be of insignificant magnitude
(Fig. 82).*

When one examines the cut surface of a stump or felled tree his eye

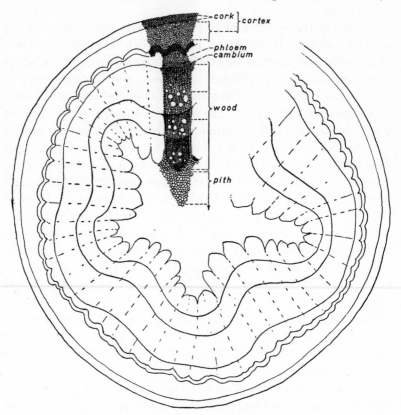

FIG. 82.—Cross-section of three-year-old oak stem

is usually caught by the concentric "annual rings." Identification of
these rings is quite possible with the naked eye, but an interpretation
of them requires microscopic examination together with some under-

---

* The fact that wood bulks far larger than does phloem in an old stem may be
attributed to two causes: (1) as noted in a recent footnote, the cambium actually produces more wood cells than phloem cells; (2) even the older wood cells,
endowed with their thick, resistant walls, persist almost indefinitely, while
many of the older phloem cells are crushed in connection with the expansion of
the stem.

standing of the action of the cambium. During the many years in
the life of the tree* the cambium continues to live and to work, but

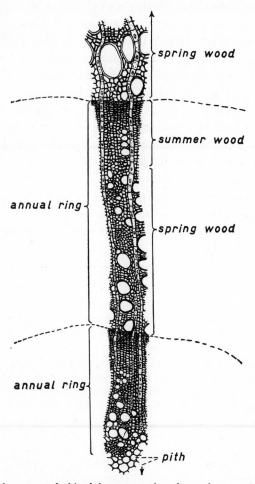

FIG. 83.—Enlargement of a bit of the cross-section of an oak stem, showing the first
and second annual rings and a portion of the third.

* Rather obviously the discussion above is limited in its application to our
"perennial" plants (trees and shrubs) in which an aerial stem is maintained for
a good many years. Most of our herbs are "annuals," with a life of only the
one growing season, though some of them maintain a perennial sporophyte
body in the form of an underground stem which sends up new shoots each year.

the quality and quantity of new wood that results depend upon the external conditions that obtain at the time. During spring and early summer when conditions (notably water supply) are at their best for the growth of the plant, the new wood cells develop into conducting tubes of large caliber. During late summer and early fall, when conditions are "on the wane," the cambium continues to lay down new wood, but this develops into cells of much smaller caliber. During winter the cambium ceases to produce new wood, and, the next spring, starts all over again with its annual program. The result is that the large-calibered cells of "spring wood" are laid down right next the small calibered cells of the "fall wood"* of the preceding year. This can be seen in detail under the microscope, and the contrast between the two types of wood is so great as to produce a line which can usually be detected quite clearly even with the naked eye (Fig. 83). Every such line or ring, therefore, represents one year in the growth of the tree, so that the age of the tree can be determined by counting the rings.†

---

* This is sometimes called "summer wood."

† In rare cases, when a period of decisive drought insinuates itself into the middle of the growing season, what are apparently two rings will be produced for a single year. The expert is able to distinguish these extra or "false rings" from the true annual rings.

If the growing season is a good one, the resulting annual ring will be wide, while a poor growing season yields a narrow annual ring. An examination of the cross-section of an old tree will, therefore, tell us something of the variation in climate through the past few centuries in the locality in which that tree was growing. Since good and bad seasons have occurred more or less at random through the past, each ten-year period has left its own distinctive record in the tree rings. Thus, if the period from 1090 A.D. to 1100 A.D. is recorded by a peculiar sequence of large and small rings in the wood of one tree, all other trees of that locality will be found to yield that same distinctive sequence for that time period. The same sequence will appear in timber that was felled long ago and incorporated into ancient buildings. Archaeologists are coming more and more to take advantage of this phenomenon for the dating of ancient structures.

The entire discussion to date has been based upon that type of stem in which the vascular elements have been arranged in one great "vascular cylinder," with phloem, cambium, and wood in concentric zones. Vascular cylinders of this general type characterize all the cone-bearing seed plants. They are also present among one of the two great groups of flowering plants (i.e., among the dicotyle-

When one examines the anatomy of the root he finds a structural arrangement that is very similar to that of the stem. In the cross-section of a *young root* one encounters on the outside the character-

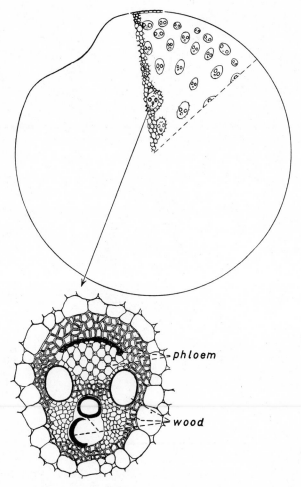

FIG. 84.—The stem of the monocotyledon. Above, cross-section of entire corn stem, showing scattered vascular bundles. Below, enlargement of a single vascular bundle.

dons). This is the group of flowering plants that includes practically all the flowering trees and shrubs that grow in this part of the world. The other group of flowering plants (monocotyledons) is composed almost entirely of herbaceous

istic epidermal layer that covers all the surfaces of all the higher plants. As stated earlier, this particular epidermis is neither thick-walled nor waterproofed, and, if the cross-section be cut at the ap-

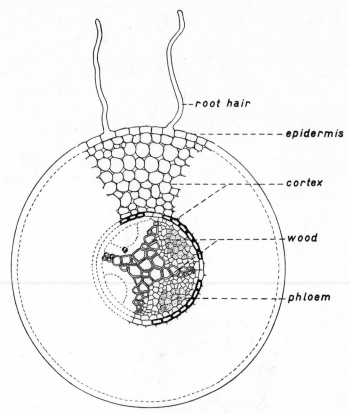

FIG. 85.—Cross-section of a root in the root-hair zone

types (though there are some perennial tree forms of monocotyledons that grow in other parts of the world). The stem of the monocotyledon characteristically has no vascular cylinder and no cambium. A few wood vessels together with a few phloem vessels are organized into a small "vascular bundle," and many such vascular bundles are scattered through the cross-section of the stem, surrounded by a general matrix of rather unspecialized tissue which partakes of the nature of both cortex and pith (Fig. 84). Lacking a cambium, these monocotyledon stems show little or no increase in diameter, save in the earlier stages of life.

propriate level, the occasional epidermal cell will be seen to protrude in the form of a long root hair. Next within comes cortex, as in the stem. Still nearer the center lie the phloem, young cambium, and wood, though the arrangement of these three parts is somewhat different from that in the young stem. Pith is usually not present (Fig. 85).

In an *old root* the resemblance to the stem is even greater. Epidermis has disappeared, cork cambium and ordinary cambium have produced much secondary tissue, and the result is successive concentric zones of cork, cortex, phloem, cambium, and wood, with only the pith lacking. This common arrangement of the tissues of old roots and old stems makes for a simple, direct connection of vascular elements at the point where root leaves off and stem begins.

When it comes to the leaves, however, one encounters greater structural differences. The characteristic leaf structure will be described later in connection with a discussion of the characteristic leaf function.

CHAPTER XIV

PHOTOSYNTHESIS AND RESPIRATION

THE theoretical viewpoint that characterizes modern natural science takes much of its flavor from two fundamental principles which have been revealed by the investigations of the physical scientist. The principle of "the conservation of matter" tells us that, though matter may pass from one form into another, it is never destroyed or created anew. The principle of "the conservation of energy" tells us that, though energy may pass from one form into another, it is never destroyed or created anew.* It is in the light of these principles that the physical scientist explains the origin, the nature, and the characteristic action of the non-living objects which make up our world, both those which were produced by nature alone and those which were artificially assembled by man.

For a long time it was felt that these two laws were inapplicable to living organisms. Even today the doubt persists, and a perennial controversy goes on between the "mechanists" and the "vitalists." According to the mechanistic view of life, the living organism is a "machine," and one may hope to explain all the activities of the living organism in terms of the (known or knowable) laws of physics and chemistry. According to the "vitalistic" view, the living organism is something distinctly more than a machine, and there are at least some of the very important activities of the organism that can never be explained in mechanistic terms, for some of the phenomena of life transcend, and are therefore not subject to, the (known or knowable) laws of physics and chemistry.

---

* Many modern physical scientists combine the two laws into one, "the conservation of matter *and* energy." They find evidence that matter is converted into energy within the stars and that energy is reconverted into matter in the stretches of interstellar space. Thus the universe is characterized by the one grand cycle instead of two. Since the energy-matter transmutations are extraterrestrial, however, and since the present book limits itself to the affairs of the earth, we shall continue to treat energy and matter as belonging to separate cycles.

With very few exceptions the men who are making contributions to biology are mechanists. The reason for this is not difficult to see. Scientists, as a rule, have been sufficiently intelligent to avoid tasks that are clearly hopeless. If a young biologist takes the vitalistic view he concludes that the more deep-seated and important phenomena of life are beyond the reach of human analysis and so refrains from throwing himself into an investigation that is hopeless from the start. But if he is a mechanist he has the faith that his research efforts will some day be rewarded; that by finding out a great deal about the living organism he may come to find out the causes of its activities, in essentially the same way that he can understand the workings of a man-made machine. The substantial contributions to biology have been based upon the products of research. It follows, therefore, that the leading biologists of today are mechanists.

By no means everything about the living organism has been interpreted successfully in mechanistic terms. What remains to be interpreted is undoubtedly greater in amount than what has been interpreted already. Even so, a great deal of progress has been made in the last century and a half; already a great many of those activities of the organism which had once been thought to be vitalistic, inscrutable, have received clear-cut mechanistic interpretation. The progress made has been not merely theoretical but highly practical. We are in debt to the mechanists for a great deal of what goes to make up modern civilization, e.g., modern medicine, public hygiene, agricultural methods. Conceivably mechanism may turn out to be untrue, in final analysis, but there can be no doubt that it has been highly useful, and the end of this usefulness is not yet in sight.

When one attempts to assign the original authorship of any such far-reaching concept as the mechanistic concept of life to any single individual, he usually finds himself confronted by a difficult and doubtless a profitless task. Regardless of formal authorship, it was the investigations of Lavoisier in 1790 (shortly before he was guillotined) that probably carried as much weight as anything else in convincing the biological public of the essential validity of the mechanistic concept. Already scientists had seen the validity of the concept of the conservation of energy as applied to the inorganic world. It remained for Lavoisier to demonstrate that it applied equally well to the organic world.

By burning fuel the man-made machine converts all of the energy that was in that fuel into the energy of work and the energy of heat. In this process the full totality of energy appears on both sides of the equation. All of the energy that was in the fuel is transformed into work or heat. No energy is dissipated into mere nothingness; nor is there any energy that is left over (provided the burning has been complete). Lavoisier showed that the same applied to the living organism. He demonstrated that the energy released by burning a given quantity of food outside the organism was exactly equal to the amount of energy that the living organism itself could secure from this food, and that the utilization of the food by the organism was in its essence a combustion, just like the burning of fuel in a steam engine. Since that time biologists have come increasingly to think of the living organism as a machine, a machine which always requires fuel because it is always running.

Accordingly plants and animals take their place along with non-living things in the grand cycle of energy. For its work and for life itself the organism is always dependent upon some source of pre-existing energy, since energy is never created anew upon this planet. And the energy that goes into the organism is never destroyed, since every bit of it is transformed into work that is done by the organism, into heat that is dissipated from the organism, and into a chemically stored form of energy that exists in the body of the organism, not only while it is alive but after it is dead.

Plants and animals likewise take their place in the grand cycle of matter. Every bit of the matter that is in the organism has been derived from pre-existing matter in a non-living form. And all the matter that has been in the organism returns once again to a non-living form.*

With respect to energy, and with respect to that matter which plays no other rôle than that of fuel, the living organism is like a man-made machine. With respect to that matter which becomes a

---

* Obviously the return of *some* of the matter to the non-living form may be *deferred* by the process of reproduction, which perpetuates into the following generation at least a small part of the protoplasm that was in the parent.

part of the body, it is quite different, for man has never yet been able to make a machine that could grow or repair itself.

These considerations serve to point out certain questions that must be answered if we are to understand the plant or the animal. From what source and in what form does the organism get its energy? From what source and in what form does it get its matter? How does it use the energy to accomplish work and to stay alive? How does it use the matter to accomplish repair, growth, and reproduction? We shall attempt to provide partial answers for these questions. No one has as yet been able to answer them fully.*

All the energy and most of the materials which supply the bodies of living organisms in general can be traced back to the process of food manufacture in green plants.† So far as is known the fundamental process is the same in all organisms that contain chlorophyll, from the blue-green algae to the flowering plants. A single description will serve, therefore, and it would be well to work out that description in terms of the flowering plants, which constitute practically the entire food source for modern man and his domesticated animals.

The actual factories are the chloroplasts, which, in the typical flowering plant, are restricted for the most part to the leaves. The distribution of chloroplasts within the leaf cells is one that makes for the maximum light supply to the maximum number of chloroplasts. In the typical thin leaf all cells save those of the epidermis and the vascular bundles contain chloroplasts. Commonly there are a score or so of small spheroid chloroplasts per cell.

To operate, these food factories must be supplied with energy and with raw materials. The energy employed is sunlight.‡

---

* And in the absolute sense, of course, no one ever will, since man can never attain an ultimate explanation which leaves nothing further to be explained.

† Exceptions to this statement are provided by the sulphur and iron bacteria (see chap. xv) and a few others which are quite independent of green plants and of other organisms as well. Since these bacteria are of negligible magnitude, however, we can afford to disregard them in our consideration of the world's food supply.

‡ Of the various components of the solar spectrum, red light provides most of the energy. All parts of the visible spectrum contributes some energy, save

The raw materials used in the process are water and carbon dioxide. A plant that lives submerged in an aqueous medium takes in water at many points. Most flowering plants, however, live in a medium of air, and in these water intake is restricted to the roots.* This water is transferred by the vascular tissues to the chloroplast-containing cells of the leaf.

In the submerged water plant, carbon dioxide dissolved in the surrounding medium diffuses directly through the plant tissues to the chloroplasts. In the land plant the source of carbon dioxide is the surrounding atmosphere.† To visualize the transfer of carbon diox-

---

only the green, which is reflected back to our eyes instead of being absorbed like the others.

Artificial light will also serve. With continuous illumination through the twenty-four hours some plants will apparently continue to manufacture food at full speed. There are others which seem to show a "fatigue" effect. At least they will manufacture as much food with twenty or twenty-two hours' illumination as they will with twenty-four.

* One could fairly generalize that the vast bulk of flowering plants fall into two categories, those in which the leaves take in no water whatsoever, and those in which there is a detectable but usually a negligibly small amount of water intake by leaves under conditions of rain and mist. A conspicuous exception is provided by the well-known "long moss" of Florida, which is actually a flowering plant that is closely related to the pineapple. This plant is an "epiphyte," meaning that in nature it perches upon another plant. (In a state of civilization we also frequently find it perching on such things as telegraph wires.) Entirely devoid of roots, the plant absorbs and strongly retains liquid rain water through an interesting valve-like action of specialized scales on its epidermis.

There are certain other flowering plants, notably some of the orchids, which grow epiphytically in humid regions of the tropics, and depend upon water which is absorbed through their remarkable "air roots." As the name implies the roots of these perching plants have no soil connections but hang freely in the air. Their silvery white color is due to a remarkable superficial layer of dead cells which soak up and retain water from the humid atmosphere in the same manner as would a piece of blotting-paper.

† The atmosphere of modern times contains 0.03 per cent of carbon dioxide. Scant as this seems, it is enough to support the vegetation which surrounds us, for the total amount of carbon dioxide in the available atmosphere is exceedingly large. If by artificial means one increases (within limits) the amount of carbon dioxide in the surrounding atmosphere, plants will manufacture more food. It is believed that the carbon dioxide content of the atmosphere has varied ap-

ide from the atmosphere to the chloroplast one must have in mind
the structure of the leaf.

Figure 86 is a diagram which reveals the three-dimensional relationships of the structures with which we are concerned. At the top
we find the upper epidermis, an intact waterproofing layer. At the
bottom is the lower epidermis, pierced here and there by tiny breathing-pores,* a few of which are shown in the diagram. Each breathing-
pore is circumscribed by a pair of sausage-shaped "guard cells."
Though these cells are a part of the epidermal layer, they are unique

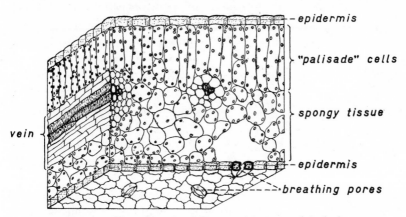

FIG. 86.—Three-dimensional diagram of a portion of the leaf

among epidermal cells in their small size, in their shape, and in containing chloroplasts (Fig. 87). Changes in shape of the guard cells in
response to changes in external conditions will increase or decrease
the size of the breathing-pore. This feature is of adaptive value in

---

preciably through geologic history, and that this has been a large factor in ac-
counting for the paucity of both plant and animal fossils that characterizes
some geologic strata and the copious fossil content of other strata.

* The technical name of the breathing-pore is "stoma" or "stomate," the
plurals being "stomata" and "stomates," respectively.

In leaves which usually have a horizontal orientation in the medium of air,
all or most of the stomata are on the lower surface. In leaves which usually
have a vertical or near vertical orientation (e.g., grasses), there are many stomata on both surfaces. In floating leaves like those of the water-lily, stomata
are restricted to the upper surface.

closing the portals of escape for water vapor when the dryness of the surrounding air endangers the welfare of the plant.*

Between the two epidermal layers there is an occasional vein, but the bulk of the tissue is made up of cells that contain chloroplasts. Those which lie just below the upper epidermis, in the case of the typical horizontal leaf, occur in a relatively compact, "palisade" arrangement, while those lower down are in a much looser, spongy arrangement which permits a free circulation of gases through the intercellular spaces. Each breathing-pore opens into an irregular air chamber, which, in turn, communicates with this system of inter-cellular spaces.

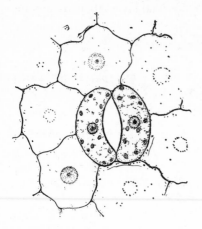

FIG. 87.—Surface view of a breathing-pore of a vascular plant, showing "guard cells" and the neighboring cells of the ordinary epidermis.

Simple diffusion (i.e., movement from a point of greater to a point of lesser concentration) accounts for the entrance of carbon dioxide gas through a breathing-pore into the air chamber. From this point the carbon dioxide continues to diffuse as a gas through the inter-cellular spaces, until at last it strikes the moist surface of one of the living cells and there goes into solution in the water. In solution it then diffuses through wall and cytoplasm to one of the chloroplasts, where, if it happens to be daytime, the carbon dioxide will be put to immediate use in the process of food manufacture.

Food manufacture by the green plant is known technically as "photosynthesis," meaning "synthesis with the assistance of light." The combination of carbon dioxide and water to form food actually involves quite a series of rapid chemical transformations. For our

---

* The degree of efficiency shown in this adaptive adjustment of the size of the breathing-pores varies considerably among different species of plants.

purpose, however, the process may be summarized by the following comparatively simple chemical equation:

$$6CO_2 + 6H_2O + \text{energy} \rightarrow C_6H_{12}O_6 + 6O_2.$$

Even without a knowledge of chemistry the student can comprehend the significant features of this equation. As in all chemical equations, this one designates merely the proportions in which the various atoms and molecules react, not the absolute quantities. Reading from left to right, the first term designates six molecules of carbon dioxide, each one of which consists of one atom of carbon in combination with two atoms of oxygen. The second term similarly refers to six molecules of water, each of which contains two atoms of hydrogen in union with one of oxygen. The third term refers not to matter but to energy, the energy that is provided by sunlight. The horizontal arrow means that those things on the left are being transformed to those on the right. $C_6H_{12}O_6$ is the formula of "glucose," one of the simple sugars.* The equation tells us that one molecule of glucose is produced, and that this molecule consists of a combination of six atoms of carbon with twelve of hydrogen and six of oxygen. The final term of the equation designates six molecules of oxygen gas. In the form of the free gas, such as occurs in our atmosphere, each oxygen molecule consists of a union of two atoms of oxygen. Summarizing: Photosynthesis involves a combination of six molecules of carbon dioxide with six of water and with the energy of sunlight, which yields one molecule of glucose as a main product and six molecules of oxygen as a by-product. (The six molecules of oxygen will diffuse out into the atmosphere unless they are used in the process of "respiration" which will be described shortly.)

In our simple equation energy is designated on the left but not on

---

* Among foods in general, the biologist recognizes three great categories: carbohydrates, fats, and proteins. Carbohydrates are composed of only the three chemical elements, carbon, hydrogen, and oxygen, the hydrogen and oxygen being usually in the same 2:1 proportion as in water. Carbohydrates, in turn, contain many subcategories, of which one of the commonest is that of the sugars. Of the many known sugars, glucose is the one that plays the most prominent rôle in the economy of plant and animal bodies.

the right. This does not mean, however, that it has been lost. Instead it is incorporated, in the form of chemical energy, into the molecule of glucose. Glucose may be thought of as the primary food, that from which all other foods are built. All foods contain energy, that same energy which was incorporated into the glucose in the process of photosynthesis. Since all living things depend upon photosynthesis by green plants, it follows that every increment of growth, however tiny, and every active movement, however slight, and every second of maintenance of life itself throughout the countless billions of plants and animals on the earth is made possible only by the energy of sunlight.

The glucose molecules that result from photosynthesis—and unthinkably large numbers of such molecules are produced during every second of illumination in every working chloroplast—exist in a state of solution in the water which suffuses the chloroplast and the surrounding cytoplasm. Some of the glucose is shortly put to use in the economy of the cell in which it was manufactured. More of it, however, diffuses out through the wall and through a few adjoining cells until it reaches the phloem tubes that are a part of the nearest vein. Within these tubes it may travel a shorter or longer distance, traversing leaf, stem, and even root; for the phloem tubes deliver glucose to the tissues at all levels of the body.*

---

* The *fact* that the phloem tubes are the channels through which the long-distance transfers of food occur is amply established by experiment. The *nature* of the transfer is still in some doubt. It is apparently more rapid than could be accounted for by simple diffusion, yet slower than the (apparently) direct flow of water (and salts) through the wood vessels. The phloem tubes consist of a series of *living* cells, and it has been suggested that the movement of glucose through the phloem is a diffusion assisted by the "streaming" or circulation of protoplasm within the phloem cells.

In contrast, the upward movement of water through the vessels of the wood is a relatively rapid mass movement. The living cells of the leaf suffer a fairly continuous loss of water in two ways: (1) some water is used in the process of photosynthesis; (2) a great deal of water is usually lost to the air by evaporation (commonly called "transpiration" in the case of plants) through those same exposed surfaces that are utilized for the entrance of carbon dioxide. In response to this depletion of their water content, the leaf cells develop a tremendous water-absorptive power, by virtue of which they pull water (directly or indirect-

Whether glucose remains in the cell where it was produced or is moved to some other cell of the body, it will be put to one of several uses. A great part of it is used as fuel to supply the protoplasm with the energy needed to support its various activities. This release of energy is not a violent "burning," such as we see in the fire-box of a steam engine, but it is the same general phenomenon. Both are "combustions" or "oxidations," i.e., chemical changes in which oxygen is combined with the fuel in a manner to release its contained energy. In the burning with which we are familiar, the oxidation is very rapid, much energy is released in a short time, and high temperatures are generated. The type of oxidation that occurs in a living organism occurs at a much lower rate, energy is released slowly, steadily, and at no time is a high temperature generated. All cells of all living organisms, plant and animal, carry on this oxidation during every minute that they continue to live. In the vast bulk of cases the chemical reaction involved is the same in detail; in a few organisms and in a few tissues the reaction is the same in principle but different in detail. The process referred to is called "respiration," and we will confine our attention to the commoner form of respiration.*

---

ly) from the woody tubes of a neighboring vein. Retained in the form of columns by the confining walls of the wood vessels, the water manifests its surprisingly great cohesive power, so that a removal of water from the top exerts an effective upward pull on the water throughout the entire vascular system.

* The student cannot get far in biology without encountering the distinction between "aerobic" respiration and "anaerobic" respiration. Aerobic respiration, or respiration with the assistance of free oxygen, is the commoner form and the only one to be outlined in our account. Since most organisms are quite dependent upon aerobic respiration they suffocate in the absence of free oxygen. Anaerobic respiration—or "fermentation," as it is more often called—occurs in the absence of free oxygen; here some of the oxygen atoms already in the glucose molecule are applied to the oxidation of another part of the molecule. Fermentation is the prevailing method of respiration in commercial yeasts (see chap. vii) and in some bacteria. Hence these organisms thrive in the absence of free oxygen. As a temporary expedient, where a rapid supply of energy is demanded, the muscle tissues of man and other animals conduct a form of fermentation. This can be continued, however, only until "fatigue" sets in, after which the original (non-fatigued) status of the tissues can be regained only through the ordinary process of respiration, based on free oxygen.

The common form of respiration involves the following chemical reaction, which amounts to an item-for-item reversal of that of photosynthesis: $C_6H_{12}C_6 + 6O_2 \rightarrow energy + 6CO_2 + 6H_2O$. Free oxygen combines with glucose in such manner as to release the energy that is contained within this fuel, and in so doing breaks down the glucose into carbon dioxide and water. Respiration of this sort is a thoroughgoing, complete form of oxidation which gets out of the fuel all of the energy that is to be had. In other words, carbon dioxide and water, the material products of respiration, retain no energy that is available for living organisms. Accordingly, these two substances may be thought of as the "ultimate wastes" of living organisms. It is apparent, therefore, that photosynthesis and respiration are compensatory processes; together they provide for a cycle of matter and energy, through which the same atoms of matter (but not the same energy) may pass repeatedly.*

Glucose is of value to the plant not only in the energy it yields but also in providing most of the material that is incorporated into the structure of the growing body. For structural purposes glucose is transformed either into more complex carbohydrates or into still other substances. In the typical plant a large part of the body consists of lifeless cell walls. Cellulose, the characteristic cell-wall sub-

---

* Aerobic respiration is a perfectly "efficient" process in that it extracts all of the energy that was in the fuel (except, of course, "sub-atomic" energy, which, so far as is known, is never used by living organisms). Fermentation (anaerobic respiration) is much less efficient. It releases only a small fraction of the energy of the glucose molecule, and yields material products which still embody energy. Thus fermentation by yeast yields alcohol, and other fermentations yield various organic acids. These several products are wastes, but not "ultimate" wastes, for they still contain energy that can be released by "further oxidation" at the hands of various organisms, notably some of the bacteria.

The fuel which provides for aerobic respiration in the green plant or in various dependent organisms which exploit that plant may be those very glucose molecules that were the product of photosynthesis and persisted in that state from the time that they were produced until the time that they were used as fuel for respiration. In many cases, however, glucose molecules may enter into the makeup of more complex carbohydrates, of fats, or of proteins, and these substances may later be used as fuel for respiration. In any event the energy can be traced to the same ultimate source.

stance, is a complex carbohydrate. The cellulose molecule is a very large one and is produced by the union of a great many glucose molecules.* The essential material that composes wood (i.e., the walls of the wood vessels) is a modified cellulose.

Fats also play some part in the structure of the plant body, and a larger part in the structure of most animal bodies. Fats are apparently an essential part of most protoplasmic membranes. Like carbohydrates, fats are composed of only the three elements, carbon, hydrogen, and oxygen, but in a different proportion and arrangement. In the plant body fats are produced by transformations of carbohydrates, and hence are also derivatives of the original glucose.

With fats and carbohydrates alone, however, there could be no protoplasm. Though protoplasm is full of mysteries as yet unsolved by the biologist, he is at least assured that the most conspicuous components of protoplasm are the proteins. Protein molecules are exceedingly large and complex. In addition to carbon, hydrogen, and oxygen, they always contain nitrogen, usually sulphur, and often phosphorus as well. Hence it is impossible for the plant to construct its proteins by a mere transformation of carbohydrates. Even though the other elements constitute only a small part of the protein molecule, that small part is quite essential. For "protein synthesis," therefore, the plant must be provided with nitrogen, sulphur, and phosphorus. These elements occur in the form of various salts in the soil.† Going into solution in the soil water, the salts diffuse into the roots, are carried upward through the conducting tubes of the wood,

---

* In this union a certain amount of H_2O is lost, so that the proportions of C, H, and O are not quite the same in cellulose as in glucose.

† In most *natural* soils there is an adequate supply of these three elements, but this is not always the case with cultivated soils. By growing and removing a succession of crops the farmer may at last reduce the local supply of one of these essential elements below the amount needed to support a healthy plant growth. This happens more commonly for nitrogen than for sulphur and phosphorus. Nitrogen, in a form available for green plants, may be replenished by manuring or with the assistance of the "nitrogen-fixing" bacteria (see chap. xv). Shortage of sulphur or phosphorus is corrected by the application of chemical fertilizers.

and diffuse out again to the various living cells of the body. Though photosynthesis can go on only in the green parts of the plant and only during the daytime, it appears that protein synthesis can be conducted by any living plant cell at any time. The cells combine glucose with nitrogen, sulphur, and phosphorus to build up the complex protein molecules, and most of these are at once transformed into new protoplasm.*

In the body of the green plant, then, the glucose may be used in respiration or in growth. It may also be disposed of in a third way. Nature has endowed plants with an unconscious thriftiness, the tendency to store up some food against a future need. This is obviously of adaptive value, for it serves to tide the plant over a period of hard conditions when photosynthesis would be difficult or impossible.

Glucose is not only the most serviceable form of fuel for plant and animal respiration, but it is also the most efficient form in which food can be transported from one part of the body to another. Not only is it highly soluble in water, the medium of food transport, but it readily diffuses through cell membranes and cytoplasm, due to the fact that its molecule is smaller than that of most other foods. When it comes to food storage, however, glucose is inferior. More energy can be stored in a small space if the glucose is transformed into some more "compact" form of food.

To a very limited extent the green plant stores food in the form of protein. This occurs to a variable degree in seeds,† but is usually of

---

* The fact that animals are dependent on the photosynthesis of green plants is probably as generally emphasized as is any principle of biology. The fact that animals are also dependent upon the protein synthesis of plants is often overlooked. The animal body has the power to transform carbohydrates to fats and the reverse, but from neither of these can it build its own proteins. For that purpose it is quite dependent upon plant proteins. In digestion the animal breaks down the large molecules of plant proteins into smaller components (the "amino-acids"), and later reassembles these components in new combinations to build proteins of the type that are characteristic of the body of that particular animal.

† Notably those of peas and beans.

very slight importance in the economy of the plant body. Nor does the animal body store much food in the form of protein.

In one respect fats are the best of storage forms; gram for gram, fat carries more energy than do carbohydrates and proteins. Even so, plants make little use of fats for storage purposes, save in the case of some seeds, such as the castor bean, which is the source of the justly famous castor oil. In general, animals go in rather extensively for fat storage. Not only is fat a reserve food supply for animals, but in the "warm-blooded" animals its distribution near the body surface provides a blanket against the loss of heat.

The favorite storage form in plants is starch, a complex carbohydrate that may be stored very compactly. In animals, too, the primary food reserve is a form of starch that is very similar to plant starch.

In the unicellular plant, starch storage must occur somewhere within the confines of the very cell that has conducted the photosynthesis. Commonly one or more spheroid accumulations of starch occur imbedded right in the chloroplast. The same thing occurs to a more limited extent in the flowering plant, but here the many-celled body makes possible the development of special storage tissues. Hence we find that, though some food stores are distributed rather diffusely through most cells of the body, there is usually some one region in which the bulk of stored food is concentrated.

Considerable storage of food occurs in the central or pith region of some upright stems. The cane sugar of commerce is derived from this portion of the sugar-cane plant, and the corresponding tissues of various palms are used extensively as a food source by the natives of tropical regions. In such cultivated plants as kohlrabi, cauliflower, and broccoli there is a moderate amount of food storage in those aerial stem parts that we eat. Many flowering plants, notably the ordinary grasses, store food in horizontal underground stems. By this device part of the product of the photosynthesis of one crop of leaves is saved to support the rapid development of a new crop of leaves. We have already noted this same arrangement among ferns (and horsetails). The Irish potato of cultivation is a prodigy of starch storage. Though the things that we call the "potatoes" are underground parts of the potato plant, they are not roots but specialized underground branches from the stem that are known technically as "tubers" (Fig. 22).

Quite a number of roots are conspicuously enlarged, "fleshy." In these cases, in addition to carrying on their ordinary functions the roots devote a large part of their tissues to the storage of food. Common illustrations of this phenomenon appear in beets, carrots, turnips, parsnips, and sweet potatoes (Fig. 78A).

There is usually not much concentration of stored food in the leaf. It is not primarily for their fuel value that man eats such things as lettuce, spinach, and celery. Where the leaves are more fleshy, as in cabbage and artichoke, the fuel value is somewhat higher. Still higher in stored food are the modified leaves of bulbs, such as the onion which man eats, and the tulip and hyacinth which he does not eat.

Whatever may be the nature and extent of food storage in the vegetative parts, the Spermatophyte may be depended upon to endow its seeds with stored food in a highly concentrated form. Beyond this, some flowering plants divert quite a bit of food to their fruits. In these last two cases, of course, the food storage is not of value in tiding the individual plant itself over a period of hard conditions, but rather in giving the young a good start in life and getting them widely distributed.*

---

* In the first chapter of this book it was pointed out that the most primitive of the three great types of reproduction was vegetative multiplication, and that this was the only type of reproduction present in such a primitive group as the blue-green algae. The point was also made that most of the higher plants, while introducing new and more specialized methods of reproduction, retained (or, perhaps, regained) that of vegetative multiplication. Among flowering plants most of the vegetative tissues have apparently become so specialized as to have lost the power to produce new individuals (save under highly exceptional circumstances). There are many species, however, in which some of the tissues, if merely placed in moist soil or other similar conditions, will develop into new and independent sporophyte bodies.

In some species such vegetative multiplication may be accomplished by

The tendency of many plants to concentrate their food stores has played a large part in molding the anatomy, physiology, habits, and general culture of man. Grazing animals have a chewing and digesting apparatus that is structurally and functionally adjusted to handling large quantities of plant material which provide much cellulose but little starch and sugar. Man could not survive on such a diet. Instead he exploits the special storage depots of plants that he finds in their seeds, fruits, and other parts, and, to make matters easier for himself, breeds new varieties of plants in which the food storage is more plentiful, more accessible, and in a more palatable form.

We have seen, then, that the disposal of glucose in the living plant falls into the three channels of respiration, growth (and repair), and

---

pieces of the root, or even of the leaf, but for the most part these powers are restricted to stems. As would be expected, the new plants are derived from the nodes of the stem, not from the internodes.

In a state of nature this form of reproduction seldom leads to any extensive distribution of the species; for that function the species is dependent upon its seeds. It does lead, however, to a more thoroughgoing exploitation of the terrain in the immediate vicinity of the parent-plant and even to a slow spread of the population. In some plants underground stems ("rhizomes" or "rootstocks") or stems which grow horizontally over the surface of the soil ("runners"), will extend laterally, and, from many of their nodes, give rise to a new set of roots below and aerial parts above. With the death of the old internodal tissues, these new plants are no longer organically connected, but provide a rather compact local population of many individuals.

The most efficient form of propagating many cultivated plants is by vegetative multiplication. Fruit trees and berry bushes are propagated by cuttings from the stems, many of our favorite flowers by their bulbs, and potatoes through the use of the tubers (on which the "eyes" are nodes which will yield new plants). A conspicuous advantage of this procedure lies in the rapidity with which sizable new plants are secured by this method. This depends upon the fact that the detached piece of vegetative tissue employed is larger and contains more nourishment than does the seed. The potato seed, for example, is tiny, and will produce only a puny, slow-growing sporophyte; while the tuber (or a fair-sized piece of the tuber) carries much starch and will yield a large vigorous, rapid-growing plant.

food storage. But what becomes of all this material and energy when the plant dies?

Some passes into the bodies of offspring, through the instrumen-

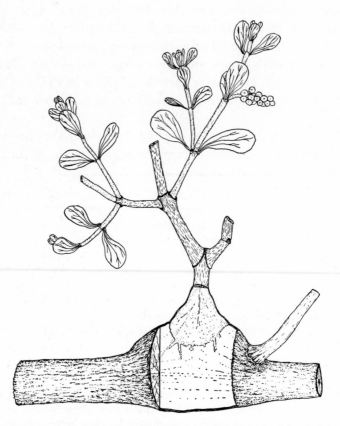

Fig. 88.—A mistletoe plant bedded in a branch of its host, shown in partial section to reveal the intimate nature of the connection.

tality of the seeds and the gametes themselves. For most seed plants this would represent only a small fraction of the total material.

If the plant merely dies and falls to the ground, we note that it does not persist indefinitely but disintegrates and adds its substance to the soil.* This disintegration is not spontaneous, but is caused by the action of the bacteria of decomposition, whose activities will be more fully outlined in the next chapter. By returning to the soil (and to the air) the matter that has been locked into the bodies of

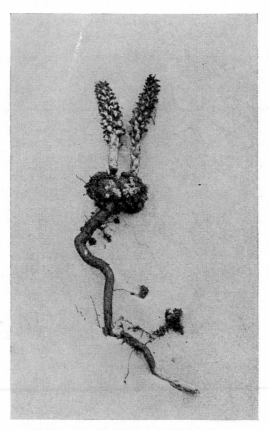

Fig. 89.—The broom-rape attached to the root of its host (which has been dissected out of the ground and thrown on a background of sand).

---

* The typical "rich" soil of forest or prairie is a mixture of inorganic particles that have been derived from the disintegration of rocks and organic particles that have been derived from the bodies of plants and animals through innumerable generations.

plants and animals, these bacteria play an important rôle in maintaining the cycle of matter in the organic world.

Under natural conditions, however, by no means all of the green plants meet their fate in this comparatively peaceful manner. The majority are directly exploited by dependent organisms of larger size than the bacteria. To some extent the fungi, and to a greater ex-

Fig. 90.—The Indian pipe

tent the animals incorporate the substance of the green plants into their own bodies. The selfsame carbon atoms will often, in fact, exist successively in the tissues of a herbivorous animal and those of a long series of carnivorous animals, until at last the final carnivorous form dies without being devoured. But the bacteria of decomposition almost always take their toll in the end, for it is they that at last break up the animal bodies and put their materials into circulation again.

There have, however, been situations in which the bacteria of de-

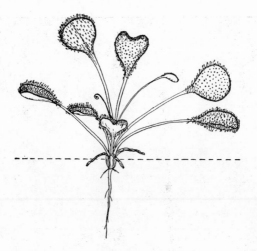

Fig. 91.—The sundew

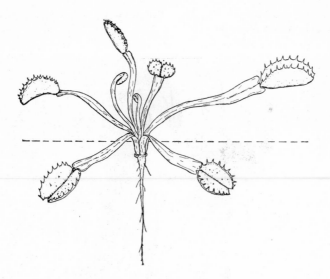

Fig. 92.—The Venus fly-trap

Fig. 93.—(See also p. 179.) Three types of pitcher plants. (Two of the photographs were taken by Walter M. Buswell.)

composition have been balked.* During the times of the "coal measures" there was an accumulation of the dead bodies of many large plants in swamp waters which inhibited bacterial action. Time and the pressure of sediments which accumulated above gradually transformed this organic material into the relatively pure form of carbon which constitutes our coal of today. Hence the energy which man releases by burning coal in modern times can be traced to the sunlight of several hundred million years ago, when it was locked up by the photosynthetic activity of those ancient plants. Most of our high-grade coal is the transformed bodies of Pteridophytes and cone-bearing seed plants, having been formed long before the flowering plants had been evolved. How much prospective coal is in the process of formation on the earth today is a question that the writer is not competent to answer.†

---

* The "embalming" of human remains serves to balk the bacteria of decomposition, at least temporarily. In terms of the amount of material involved, however, this practice is a factor of negligible magnitude in the organic world as a whole, and of course it has been in operation for only a negligibly small fraction of the total history of life.

† From the foregoing descriptions the student has gathered the impression that all vascular plants are endowed with a machinery for photosynthesis which makes them quite independent of other organisms. Actually, this proposition must be qualified by exceptions of two types.

1. So far as is known, the higher green plants are indirectly dependent for their nitrogen supply upon the activities of certain groups of bacteria. This matter will be explained in the chapter on bacteria.

2. A very few of the flowering plants have adopted unusual nutritive relationships that are characterized by a partial or complete dependence upon other organisms:

The well-known "mistletoe" might be called a "partial parasite." It becomes firmly attached in a lofty position on the trunk or branch of some tree (e.g., pine, oak) by means of strange root-like processes which push themselves into the vascular cylinder of the host. Though the mistletoe conducts its own photosynthesis, it is quite dependent on its host for water and soil salts (Fig. 88).

The "broom-rape" is a complete parasite. Devoid of chlorophyll, it attaches to the roots of various hosts and derives food from that source (Fig. 89). An even commoner example of this type of thing is provided by the "dodder."

The picturesque "Indian pipe," on the other hand, is a saprophyte. Its ghostly white body is nourished from the decaying wood and leaves that accumulate in moist forests (Fig. 90).

Another group with weird nutritive relationships is that of the "insectivorous plants," whose capacities for devouring animals have often been exaggerated in popular fiction. The common "sundew" captures small insects on the "flypaper principle" through exudation of a sticky substance from glandular hairs on the leaf (Fig. 91). The "Venus fly-trap" has a "hinged," trap-shaped leaf which actually snaps shut upon the unfortunate fly that touches one of the sensitive ("trigger-like") hairs on the leaf surface (Fig. 92). The several "pitcher plants"—fairly common in California and the Gulf states, and found to a more limited extent in the Great Lakes region—have funnel-shaped leaves in which water collects. Insects, falling into these reservoirs, encounter structural features that make it next to impossible for them to extricate themselves (Fig. 93). In all cases there is a secretion of enzymes which digest the bodies of the captured insects and thus provide a certain amount of nitrogenous food that is absorbed by the plant. But in all cases the plant is green and supplies the bulk of its nutritive needs by conducting photosynthesis.

CHAPTER XV

BACTERIA

W E WOULD do well, at this point, to give some attention to a group of plants, which, on the basis of their relationships, belong much earlier in our story. Bacteria are regarded as plants by most biologists, largely because of their apparent similarity to the blue-green algae. We have deferred a discussion of them until the present time in order to take them up in a context which will clarify the important rôles that they play in those cycles of matter and energy that unify the organic world.

As a group, the bacteria are ancient—perhaps even more so than the blue-green algae. Those forms which seem most important to man, however, were probably evolved rather recently in geologic time, for they are adapted to interactions with various living organisms that are themselves relatively recent products of evolution.

Like the blue-green algae, bacteria are all unicellular. Many of them, however, like most of the blue-greens, tend to form colonies— most commonly filamentous colonies. Like the blue-green algae, they all possess a rather generalized protoplasm, with no clean-cut differentiation of cell organs. Many, but by no means all, are able to swim through a liquid medium with the assistance of minute cilia. As in the blue-green algae, the only method of reproduction is by simple cell division. In rate of reproduction, however, many of them excel the best of the blue-greens. Under favorable conditions some bacteria may grow to full size and divide every twenty minutes, and it is largely this feature that accounts for the appearance of serious and large-scale symptoms in a host that has been infected a few days or a few hours before.*

---

* The student might amuse himself by calculating the number of descendants that would result from a single ancestral bacterium in 24 hours, on the assumptions that cell division occurred every 20 minutes and that all of the offspring survived.

Bacteria are the smallest known living organisms.* Decidedly smaller even than the blue-green algae, they range in size down to the limits of visibility under our best microscopes. Roughly speaking, an average sized bacterium would be about 1/25,000 of an inch in its shortest dimension and two or more times that in its longest.

In shape of cell there is more variation than among the usually spherical blue-greens. Spherical bacteria are spoken of as the "coccus" forms; rod-shaped bacteria, as the "bacillus" forms; and banana- or corkscrew-shaped bacteria as the "spirillum" forms (Fig. 94). Thousands of species are known; their classification takes cog-

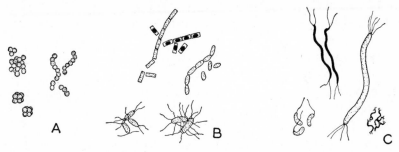

FIG. 94.—An assortment of bacteria. *A*, coccus forms. *B*, bacillus forms; "spore" formation is exhibited by one of these specimens. *C*, spirillum forms.

nizance not only of shape, size, and the types of colonies formed, but of the reactions produced in non-living nutritive media and in living hosts.

Bacteria are, in general, even more resistant to extremes of conditions than are the blue-greens. Even in their active vegetative state, many of them are able to withstand degrees of heat, cold, and acidity that are fatal to most organisms. Beyond this there are some of the rod-shaped forms that can strikingly increase their powers of resistance by forming the so-called "spores." The spore of the bacteriologist is not an agent of reproduction but merely a dormant condition of the ordinary cell. In spore formation the protoplasm rounds up more compactly than before and usually surrounds itself with a thick wall, within which it may remain dormant but viable for a pro-

---

* That is, if we omit the filterable viruses (see footnote on p. 58) and confine our consideration to the organisms we can be more sure of.

digious length of time (Fig. 94B). In this state some bacteria have been known to withstand extreme dryness for many years, to re-awaken and return to the active vegetative state when conditions of moisture and warmth are provided.* Many bacteria will resist long freezing in ice, and this includes the typhoid bacillus. At the other extreme are some which, as spores, will resist boiling for several hours. For complete sterilization, therefore, there must be boiling under pressure, or the thoroughgoing application of strong germicid-al fluids (antiseptics) to every spot in which bacteria might be lurk-ing.†

In view of these record-breaking powers of resistance, it is not sur-prising that bacteria also hold the record for ubiquity. All ordinary soils are teeming with bacteria, particularly those upper layers of the soil that are rich in organic matter. Natural streams and sea water contain many, while small lakes, ponds, and puddles contain many more. The number of bacteria present in every drop of water at popular bathing beaches is shockingly high. The air is full of them. In the comparatively pure air of the mountain tops they are quite rare, but in the air of crowded cities the bacterial population is enormous. This appears to be due in part to the fact that bacteria "ride the dust particles" in most of their sustained journeys through the air.

Bacteria are almost always present on and in the bodies of other organisms. Man is no exception. On his skin, in his mouth, and par-

---

* Again the student who enjoys calculations might attack the following prob-lem: Assuming twenty minutes for the normal length of a bacterium generation, and twenty years for the maximum period of dormancy, during how many bacterial generations did this Rip Van Winkle of bacteria sleep? Then, assum-ing that Rip was a man, but endowed with the capacity to remain dormant for as many generations as did this bacterium, when did he first go to sleep to wake up in the year 1900 A.D.?

† As might be suspected, the strength of the germicide needed to kill bac-teria in their spore forms is ten or more times what is needed to kill them in their ordinary vegetative phase. It is a very fortunate thing for man that most of the bacteria responsible for his more serious diseases are incapable of forming spores. Were this not the case, the continuous warfare that has been going on between man and the bacteria might well have ended long ago in overwhelming victory for the latter.

ticularly throughout the lower stretches of his alimentary tract he
harbors millions of bacteria. These intimate companions of man are
not at all harmful under ordinary circumstances, but may become
so the minute that there is any serious upset in the structural and
functional defenses with which nature has provided man's body.

In most of the features cited we see a resemblance between bacteria and the blue-green algae. The big, consistent difference lies in
the absence of chlorophyll from the bacteria. As a result most of
them are dependent organisms. In working out their nutritive relationships, we shall see how some bacteria have become man's most
dangerous enemies, while others are among his most important bene-
factors.

A very few bacteria are independent.* A footnote in an earlier
chapter pointed out that some bacteria can conduct a form of photosynthesis on the basis of pigments other than chlorophyll, and that a
few others can synthesize carbohydrates even in the absence of sunlight, on the basis of energy derived from the oxidation of iron. On
the basis of both structure and physiology these appear to be the
simplest of living things, but whether they represent an approxima-
tion of the original ancestral condition or the product of retrogressive
evolution remains an open question.

The vast majority of bacteria are dependent, and the energy
stores exploited by some of them are such as to carry them into the
field of man's interests.

A good many bacteria parasitize both plants and animals, including man's cultivated plants, man's domesticated animals, and man
himself. Infection of a host depends upon the kind of bacteria and
the kind of host, and upon whether the existing conditions favor the
defenses of the host or the penetrating power of the bacteria. Some
bacteria enter man's body through tiny scratches in his skin that he
has not taken the trouble to disinfect, some through his alimentary
tract, some through his respiratory tract. Even when they have entered, the bacterial hordes are, more often than not, defeated by the
various defense mechanisms with which man's body is provided.

---

* Technically the independent bacteria are known as the "autotrophic"
forms and the dependent bacteria as the "heterotrophic" forms.

But there are many humans in which the defenses against certain bacteria are either permanently or temporarily weak. In these cases the bacteria gain a foothold, with ultimate consequences that are more or less serious.

It is not through a simple devouring of tissues that bacteria work their greatest havoc. It is more through the poisons that they liberate as a by-product of their own life-processes. In some the poisons are in significant concentration only in the immediate locality occupied by bacteria. There results an abscess or local destruction of host tissues. In others, powerful poisons are produced and pass into man's circulatory system. Fever results, and a weakening of the host at many points. Further description of the nature of bacterial disease is obviously beyond the scope of this small book on plants.

In addition to this destructive attack upon the body of man himself and upon those other organisms in which he is interested, bacteria (together with some of the fungi) are responsible for the spoilage of his foods. It is a far-reaching principle of chemistry that the rate of a reaction increases with an increase in temperature. Since life-processes are in large part merely chemical reactions, it follows that the rate of the life-processes will increase with an increase in temperature.* At low temperatures bacteria may continue to live, but they grow and reproduce slowly if at all. Since bacteria are practically ubiquitous, some will inevitably lodge upon most of our foods. Store the foods in a warm place, and bacterial spoilage will proceed apace. Store them in refrigerators, and the spoilage may be so slow as to be negligible.

---

* This biological principle applies only within obvious limits. Raise the temperature too high and you destroy the machinery for the life-processes. For every organism there is a characteristic temperature range which it will tolerate. The "minimum" temperature for the species is that below which death results; the "maximum" is that above which death results. Somewhere between these two extremes lies a point or small zone which biologists refer to as the "optimum" temperature, for it is at this temperature that the constructive life-activities of the organism proceed at the greatest rate, so that the individual and the population grow luxuriantly. The principle of increase in rate of *constructive* life-activity with increase in temperature applies only up to the optimum temperature.

It is another principle of chemistry that the rate of many reactions is strikingly expedited by putting the reacting substances into solution. The most generally effective solvent is water, and this is the medium to which protoplasm is adjusted. For this reason, the other very effective method of reducing bacterial spoilage is to dry out ("dehydrate" or "desiccate") the food that is to be stored, and then make sure that the material is kept dry through the storage period.

The "negative economic significance" of bacteria, particularly in disease production, has been increasingly publicized since the days of Pasteur, and has gone far to clear man's understanding of the world in which he lives (see p. 58). The general public, however, little realizes that there is another side to the picture—that bacteria are man's friends as well as his enemies. The student of biology knows this and knows that it is no small matter. It would be no over-statement to say that man could not exist on the earth as it is today were it not for the useful activities of some bacteria.

This "positive economic significance" of bacteria is due largely to a limitation in the powers of green plants. For its proteins and its protoplasm the green plant must have nitrogen. Approximately 80 per cent of our atmosphere is free nitrogen gas, so that a sea of nitrogen not only bathes the aerial parts of the green plant, but also percolates through the air spaces of the soil and bathes the roots. This limitless supply of nitrogen is of no more value to the green plant than is sea water to the thirsty, shipwrecked sailor. The green plant cannot use nitrogen in this form. In general, to be available to the green plant, nitrogen must be in the form of "nitrate," in which one nitrogen atom has been combined with three atoms of oxygen.

Some nitrate is present in most soils.* Through their lives, green plants are continuously drawing upon this supply. If there is no replacement, a time will arrive when the nitrate supply is reduced below what is necessary to support plant growth. Some nitrate may be added to the soil by non-living agencies,† but this is negligibly small

---

* In the form of calcium nitrate, potassium nitrate, and magnesium nitrate.

† For example, the oxidation of small amounts of nitrogen by lightning flashes.

in amount. Were it not for living agencies, green plants would soon perish from nitrogen-starvation.

Nature works out a partial solution for this problem by a cycle

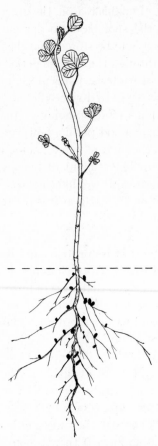

which serves to feed new generations of green plants with the same nitrogen that has already served the needs of pre-existing green plants. The dead plant or animal body gradually decomposes as it lies upon the ground. This phenomenon recurs so frequently and so copiously that most mortals have perceived it, but those who have thought about it at all have usually attributed it rather vaguely to the action of some non-living agency of nature. As a matter of fact, a corpse would break up very slowly indeed under the attack of wind, rain, and the like. Were these the only agencies of decomposition, the ground surface would today be hopelessly cluttered up with the accumulated corpses of many generations past. Actually, it is the bacteria of decomposition that take care of all this débris, gradually breaking it up and returning its component materials to the soil (and to the air). Through the action of these bacteria of decomposition, plant and animal proteins are converted into ammonia in the soil; other bacteria (the "nitrifying bacteria") then seize upon the ammonia and convert it into nitrates. These nitrates may then

FIG. 95.—Root "nodules" ("tubercles") produced by nitrogen-fixing bacteria on the roots of a clover plant.

serve the needs of new generations of green plants and of the animals that eat the plants.

Here, then, is a complete cycle which might seem to provide for the living organisms of all time to come. By itself, however, it is not a perfect solution to the problem. Through the action of rain and soil water, some nitrates are usually leached out of the soil and carried away. Worse than that, there are some bacteria, known as the "denitrifying bacteria," which act to break up soil nitrates in such manner as to return free nitrogen gas to the air. Due to these steady losses, the system would inevitably run itself down were there not some agency for the reconversion of atmospheric nitrogen into soil nitrates.

Exactly this rôle is played by the "nitrogen-fixing" bacteria. Without these humble, microscopic organisms, man would soon pass out of the picture, along with most other forms of life. Of the nitrogen-fixers there are two categories. Some exist here and there in the soil in a free state, having no intimate connection with other living organisms. More nitrogen is fixed, however, by those of the other category which live in the root "nodules" ("tubercles") of "leguminous" plants (e.g., beans, peas, clover, alfalfa). The presence of these bacteria in the roots acts as a stimulus to growth, so that the local root tissues enlarge to produce a tumor-like swelling that may be several times the diameter of the normal root (Fig. 95). It is within these nodules that the bacteria make their homes and carry on their peculiar work of nitrogen-fixing. The result is an ample supply of nitrate for the host plant.*

The farmer usually relies upon the nitrogen-fixers to rejuvenate his exhausted soils. After a series of ordinary crops has brought the local nitrate supply to a dangerously low level, the farmer may allow that field to lie fallow for a season, so that the free-living nitrogen-fixers may have opportunity to replenish the supply. If he wishes a

---

* The foregoing paragraph provides the conventional textbook account of "nitrogen-fixation," implying that the "nitrogen-fixing" bacteria themselves have the power to carry free atmospheric nitrogen way over into the state of nitrate, in which it is available to green plants. This result is indeed attained, but apparently not by the independent action of the "nitrogen-fixing" bacteria. Recent studies indicate that these bacteria convert atmospheric nitrogen into ammonia and use this as one of the materials needed to form their own proteins. Upon the death of these bacterial bodies, their proteins are converted (by the "bacteria of decomposition" and the "nitrifying bacteria") into soil nitrates, just as are the proteins of the dead bodies of other organisms.

more copious replenishment, he sows leguminous plants in the field, and at the end of the season plows their bodies back into the soil.*

One can see, therefore, that the oft-mentioned "nitrogen cycle"

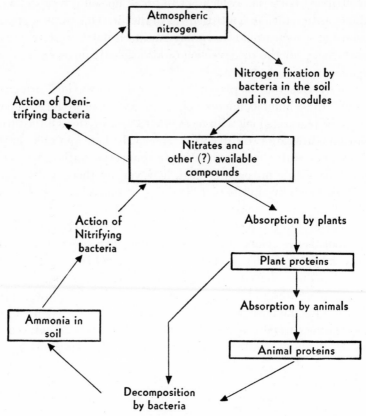

FIG. 96.—The nitrogen cycle, simplified

---

* The nitrate supply will not be replenished, and the leguminous plants will not themselves thrive, unless the proper nodule-forming bacteria are present. If these are known to infest the local soil, well and good. If not, the farmer would do well to secure (from the Department of Agriculture) cultures of the bacteria with which to inoculate the seeds before sowing, so that the bacteria will be there to enter the young roots when they appear.

Much earlier the lichens provided us with a striking example of "symbiosis," the "living together" of organisms of different types in an intimate union which benefits both. The root nodules of leguminous plants are another famous example of this phenomenon.

in the organic world is really a compound of two smaller cycles. Given nitrates in the soil, we may have a cycle which involves successively the following organisms: green plant—(animal)—bacteria of decomposition—nitrifying bacteria. In the other cycle we have the denitrifying bacteria and nitrogen-fixers continuously working at

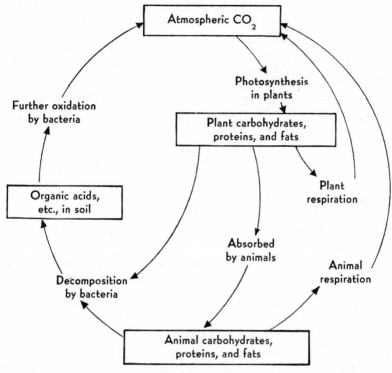

FIG. 97.—The carbon cycle, simplified

cross-purposes, the former changing nitrates into free nitrogen and the latter effecting the reverse change. The two cycles feed into one another at the point of the soil nitrates. The relationships are recorded graphically in Figure 96.

Actually there is a cycle for every one of the "essential elements" in plant and animal bodies,* and in each cycle the bacteria of de-

---

* For years, botanists have recognized "ten essential elements" for green plants. All are needed for normal plant growth, though some are required in very

composition play a rôle. Most significant, however, are the nitrogen and carbon cycles.

In large part the carbon cycle has been described already. Atmospheric carbon dioxide is incorporated by photosynthesis into plant carbohydrates. Some of these later become fats and some enter into the protein molecules. The cycle is completed without the intervention of other organisms when plant respiration returns carbon dioxide to the atmosphere.

The carbon cycle becomes longer when herbivorous animals (or other dependent organisms) insinuate themselves, and still larger when the herbivores are, in turn, devoured by carnivores. Thus, the carbon of plant carbohydrate, fat, and protein may become the carbon of animal carbohydrate, fat, and protein. Much of this carbon, in the form of carbon dioxide, is steadily poured back into the atmosphere by animal respiration.

The cycle is longer yet for those carbon atoms that remain a part of undevoured plant and animal corpses. At this point the bacteria of decomposition get in their work, breaking down the various body materials into simpler forms, such as organic acids. These organic acids do not exist for long in the soil, however. They still contain some chemically stored energy, and this energy is soon exploited by a second crew of bacteria, which oxidize the organic acids and return the carbon, in the form of carbon dioxide, to the atmosphere. Graphically, the carbon cycle is represented in Figure 97.

---

small quantities. This classic list of ten is: carbon, hydrogen, oxygen, nitrogen, sulphur, phosphorus, calcium, magnesium, potassium, and iron. There are now indications that traces of boron, manganese, zinc, and perhaps still others are also necessary.

CHAPTER XVI

PRIMITIVE SEED PLANTS

IN AN earlier chapter we pointed out certain distinctions between the two major subdivisions of the Spermatophytes, stating that the Gymnosperms, or "cone-bearers," were decidedly more primitive than the Angiosperms, or "flower-bearers." This point is strikingly established by the fossil record. Angiosperms appear first in the latter part of the Mesozoic era at a period estimated to be about a hundred and twenty million years ago. Gymnosperms, on the other hand, seem to have been fairly plentiful as early as the middle of the Paleozoic era, or at a period about three hundred and ninety million years ago. A few of the Gymnosperms, in fact, are found as early in the fossil record as are any of the Pteridophytes, and the forests of the coal measures were a mixture of Pteridophyte and Gymnosperm forms.

One probably should not infer from this that Pteridophytes and Gymnosperms were simultaneous in origin. All Gymnosperms show features sufficiently advanced as to make it almost incredible that they were derived directly from Bryophytes. It seems much more plausible that Pteridophytes were derived from Bryophytes and then, in turn, gave rise to Gymnosperm descendants. The difficulty about the fossil record doubtless lies in the fact that (1) the period of origin of Pteridophytes from Bryophytes, (2) the early period of evolution within the Pteridophyte group itself, and (3) the period of origin of the earliest Gymnosperms from Pteridophytes, all antedated the period at which a satisfactory fossil record of plants begins.* By the time that a satisfactory record begins, in the middle of the Paleozoic, plant evolution had arrived at stage (4), in which Pteridophytes were approaching their maximum display and Gym-

---

* For geologic reasons, the fossil record of the first half of the Paleozoic era is exclusively marine and tells nothing about the land flora of the time. Prepaleozoic deposits have been so metamorphosed as to obliterate the record.

nosperms had already made a start. For the next two hundred million years or more these two groups together dominated the land surface. When, at the end of this long stretch of time, the Angiosperms finally appeared, the decline in dominance of both Pteridophytes and Gymnosperms occurred with comparative rapidity.

As to the exact subdivision of Pteridophytes that gave rise to Gymnosperms, there is a difference of opinion. The Cycadofilicales, a group of Gymnosperms that was plentiful during the Paleozoic era but has long been extinct, provides, as the name implies, a beautiful transition between the ferns (Filicales) and the cycads (Cycadales), a Gymnosperm group that reached its peak in the Mesozoic era and is still represented by a few tropical species. It is by no means easy, however, to visualize how the Cycadofilicales, either directly or indirectly through the Cycads, may have given rise to all other types of seed plants. There is a distinct possibility that the great assemblage of forms which today possess seeds had a "polyphyletic" rather than "monophyletic" origin; in other words, that some seed plants may have been derived from one subdivision of the Pteridophytes, while other seed plants had independent origins in other subdivisions of the Pteridophytes. Such an idea embodies the assumption that the seed itself was evolved not merely once but on several occasions. In general, any such assumption runs counter to the bulk of evolutionary evidence. It is not, however, without precedent, and in the present instance is supported by the following consideration: The fossil record makes it clear that heterospory was evolved independently in several Pteridophyte groups, and heterospory is clearly a step toward the production of the seed.

Among living Gymnosperms, four orders are usually recognized:

The order Gnetales is a strange little group which has left practically no fossil record behind it. Apparently this represents a short evolutionary sideline, rather modern in origin as Gymnosperms go, which has become adapted to life in the arid districts of the tropics (Fig. 98).

Still smaller but much more ancient is the order Ginkgoales. With a fossil record that reaches way back into the Paleozoic and an extensive display of forms in the Mesozoic, the order is today represented by only a single species, *Ginkgo biloba*, the "maidenhair tree"

(Fig. 99). Clearly this is an archaic type. In fact it has been suggested that the group would have become extinct ere this had it not been carefully cultivated in the temple gardens of China and Japan.

FIG. 98.—Photograph of *Ephedra*, one of the few living representatives of the order Gnetales.

The order Cycadales, though not so ancient in itself, has a rather clear line of ancestry, through a series of fossil forms, back to the Cycadofilicales of the Paleozoic. Today the cycads are represented by nine genera and about a hundred species, all of which are tropical (Fig. 100).

FIG. 99.—Photograph of the "maidenhair tree" (*Ginkgo biloba*) growing on the University of Chicago campus. (The leaves of this form exhibit the forked veining that is otherwise restricted to the fern group.)

In contrast with the cycads, the conifers (order Coniferales) are for the most part restricted to the north and south temperate zones.

This is by far the most conspicuous order of living Gymnosperms in terms of number of individuals, number of species (almost 400), size

FIG. 100.—Photograph of a cycad, growing in the University of Chicago greenhouses.

of individuals (many of which are the giants of the plant world, some being trees almost 400 feet in height, e.g., giant sequoia), and economic significance (providing the "soft wood" trees of the lumbering industry). With few exceptions the conifer is an "evergreen," retaining its leaves through the winter season. Though a few conifers have broad-bladed leaves, more characteristic of the group is the narrow "needle" type of leaf.* The more familiar conifers of this part of the world are the pine, spruce, fir, hemlock, cedar, juniper, and giant sequoia.

The life-cycle of the pine is the life-cycle of conifers in general, and is fairly representative of the entire Gymnosperm group. Like all seed plants, and like *Selaginella*, the pine is heterosporous. Unlike *Selaginella*, but like all conifers and most Gymnosperms, the pine produces two types of strobili.

The one type of strobilus or cone is composed of microsporophylls only. The individual microsporophyll is small (about $\frac{1}{8}$ inch in length) and delicate in texture. Its lower surface is completely occupied by a pair of elongated microsporangia, each packed with thousands of microspores. Prior to the release of the microspores, the sporophylls are arranged with perfect compactness in the strobilus, which is about a quarter of an inch in diameter and an inch in length. The strobili are produced on the tree in clusters of about a dozen each, and the entire tree may bear thousands of these clusters. Hence the total output of microspores by a single pine tree is prodigiously large. During a brief period of a few days' duration in each season most of these tiny yellowish spores are released, producing an effect that has often been referred to as "showers of sulphur." Each microspore is equipped with a pair of (perfectly passive)"wings," by virtue of which it may be sustained in the air for a considerable time before it lodges somewhere (Fig. 101).

Since Spermatophytes are in general our largest and most conspicuous plants, it is not surprising that the earliest botanists gave most of their attention to this group. For the various reproductive struc-

---

* The needle leaf, with its small surface area in proportion to volume and its extremely heavy epidermis, has adaptive value in the reduction of water loss. Though we do not find conifers in arid regions, their evergreen habit subjects them to conditions which are similar in effect. During the winter season intake of water by the roots is much reduced, for much of the soil water is effectively tied up in the form of ice.

tures which they encountered they coined names which, of course, in no way reflected the relationship between Spermatophytes and Pteridophytes or the correspondence ("homology") between the successive stages in the two life-cycles. It was later, after the Pteridophytes had been studied intensively, that botanists produced those more enlightened terms which refer to the various structures which occur in connection with heterospory. It is these more modern terms

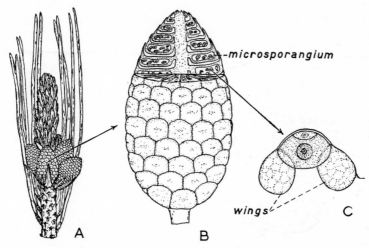

FIG. 101.—*A*, cluster of small (staminate) cones carried on a twig of the pine. *B*, a single cone, shown in partial section to reveal the nature of the microsporophylls. *C*, an enlargement of one of the winged pollen grains, showing the young male gametophyte within the old microspore wall.

that we have employed in our descriptions thus far. The older terms were used so long, however, as to intrench themselves firmly. The unfortunate result is that the student of botany must familiarize himself with both sets of terms, since botanical literature makes frequent use of both. The strobilus of our modern terminology is the "cone" of the older and more popular usage. Our microsporophyll is a "stamen"; our microsporangium, a "pollen sac"; and the microspores themselves, as we shall see, develop into the "pollen" or "pollen grains." Since there are the two types of strobili in the pine, the one we have been discussing is distinguished as a "staminate strobilus" or "staminate cone."

The other type of cone is a much larger affair. This is the "pine cone" of popular experience. The megasporophylls which make it up are not only much larger than were the microsporophylls but are much tougher in texture. On the upper surface of each, near the point of attachment of the sporophyll, is a pair of somewhat flattened megasporangia. Under the older terminology, the megasporangium was an "ovule" (Fig. 102). Here the ignorance of the older terminol-

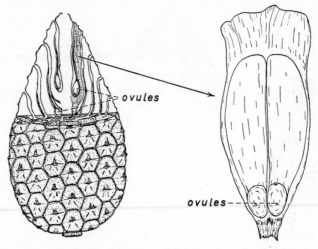

ovules

ovules --

FIG. 102.—At the left, the ovulate cone of the pine, shown in partial section. At the right, an enlargement of a single megasporophyll, seen from the upper side.

ogy is revealed clearly, for ovule means "little egg" (taken from the ovum or egg of animals). The pine does, indeed, contain eggs, but they should not be confused with the much larger megasporangia. The megasporophyll was called a "carpel"; while the entire cone was referred to as a "carpellate cone," sometimes as an "ovulate cone."

It is the megasporangium that is destined to become the seed, but not before or unless quite a program of events is properly enacted. The megasporangium or ovule of seed plants has a wall many layers of cells thick. Bedded down in the center of this tissue is a single megaspore,* which is never shed but remains in the spot where it is

---

* Actually four megaspores are at first produced, but three of them are aborted and only one develops further.

produced. The megaspore, of course, produces a female gametophyte. This is even smaller than in Selaginella and is retained completely within the megaspore wall. At one end the female gametophyte produces a very few imbedded female sex organs, each with its
large, passive egg (Fig. 103).

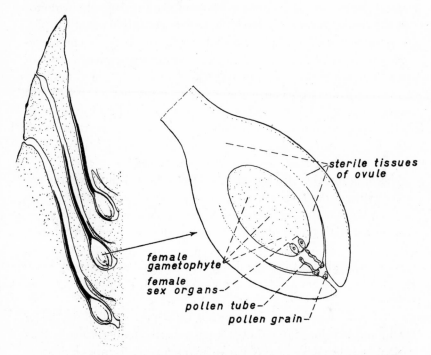

sterile tissues
of ovule

female
gametophyte

female
sex organs

pollen tube

pollen grain

Fig. 103.—At the left, longitudinal section of a portion of the ovulate cone. At the
right, enlargement of a single ovule, shown in the same orientation. Pollen grains, having slipped down and come to rest in the angle where the megasporophyll is attached to
the axis of the strobilus, are drawn into the little opening at the end of the ovule by an
interesting process. Through this opening the tissues of the ovule first secrete a gummy substance to which the pollen grains adhere. Later this gummy substance is re-
sorbed by the tissues of the ovule, and the pollen grains are drawn back along with it
through the opening.

The microspore, even before it has been shed from the microsporangium, has commenced to produce internally a diminutive
male gametophyte which consists of only a very few cells (the exact
number of cells being a variable among Gymnosperms). The pollen

grains, therefore, at the time they are being showered down from the staminate cones, are not simple microspores, but objects which consist of microscopic male gametophytes within microspore walls.

"Pollination," or the transfer of pollen from the place of its production to the place where it may function, is in the pine about as wasteful a process as occurs anywhere in plant or animal kingdoms. At the time of pollen-shedding the ovulate cones of most of the pine trees in the vicinity are prepared to "receive" pollen. The megasporophylls have pulled apart at their outer edges, leaving a series of rifts into which pollen grains might be carried. There is nothing, however, to direct the pollen grains. They are at the mercy of chance currents of air, and the ovulate cones are tiny targets in the entire field through which the pollen grains travel. It is for this reason that the production of pollen grains must be prodigious to insure that even a few of them will lodge in the ovulate cones.

Those grains that do enter the ovulate cones will slip down the upward slanted megasporophylls and come to rest in the angle where the sporophyll attaches to the axis of the strobilus. This brings them in contact with the lower ends of the ovules, at a point where the ovule walls are thinner than elsewhere. At this stage, therefore, male gametophytes within pollen grains are separated by only a few layers of cells from a female gametophyte that is imbedded within the ovule.

The male gametophyte now pushes through the wall of the pollen grain and develops into what is known as the "pollen tube." This pollen tube resembles a tiny parasitic fungus, for it grows by eating its way through those cell layers of the megasporangium wall which are covering the female gametophyte at this point. At this same period the male gametophyte "matures" by producing two sperms. These are not produced within any retaining structure which can be recognized as a male sex organ, but are merely two "free" cells that are carried along within the advancing tip of the pollen tybe. The pollen tube reaches the female gametophyte at the point where it carries the several imbedded female sex organs. Here the tip of the pollen tube bursts, releasing the two (non-ciliated) sperms, one of which usually succeeds in fertilizing an egg (Fig. 103).

Since several pollen tubes may be present and several female sex organs regularly are present, it is quite possible for more than one zygote to be produced within the same female gametophyte. Almost invariably, however, the embryo sporophyte which results from one of the zygotes will outgrow and "crowd out" the sister-embryos which may have started to develop at its side. The final result is a single embryo sporophyte developing within the old female gametophyte, which is surrounded, in turn, by the simple cell wall of the old megaspore and the many-layered wall of the megasporangium or ovule. The whole structure is still being carried on the upper surface of a megasporophyll, which is merely one of the units in an ovulate strobilus.

For reasons that are still mysteries to the botanist, fertilization, or the events which lead to it, initiate changes not merely in the resulting zygote, but also in the megasporangium wall. In this region, which was up to this time a soft tissue, cell changes occur which gradually transform a part of the megasporangium wall into a tough coat, completely surrounding the female gametophyte. When this wall is perfected it has an inevitable effect upon what lies within. The embryo sporophyte develops parasitically at the expense of the female gametophyte only so long as it receives a supply of water from conducting tissues which are feeding into the outer regions of the megasporangium. As the hard coat on the outside completes itself, it cuts off this water supply and hermetically seals off the female gametophyte and its contained embryo. As a result the embryo is thrown into a state of dormancy. The first chapter of its development is at an end, and it may now lie for months or even for many years in a state of arrested development, well protected within the surrounding structures.

What was once the ovule or megasporangium is now the seed. For Gymnosperms, then, we may define a seed as a transformed megasporangium which now consists of a tough seed coat, containing the remnants of an old female gametophyte and a young embryo sporophyte.

In the pine, as in many other conifers, a long flap of the megasporangium wall, which has been stretched outward along the upper surface of the megasporophyll, becomes a single "wing" for the seed. When the old ovulate cone gradually opens out and sheds the seeds, these "wings," like those of the pollen grains, increase greatly the chances that seeds will be carried to some distance by the wind. Biologically this is of large importance, for the seed is the only agent of distribution in an organism of this type. If pine seeds dropped to the ground like plummets from the points where they were produced, we would doubtless today find groves of pines in only a few restricted localities. By virtue of the winged seed, however, a slow but effective distribution of pines has occurred through the past, so that today we find them present over extensive areas (Fig. 104).

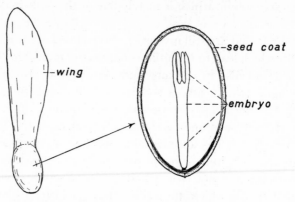

Fig. 104.—At the left, the winged seed of the pine, surface view. At the right, a longitudinal section through the pine seed, wing omitted.

In the matter of seeds the "wastefulness" of the pine may not be so great as for its pollen grains. Even so, it is but a fraction of 1 per cent of the seeds produced that ordinarily succeed in developing pine trees under the conditions most commonly encountered in nature. Carried by the wind, some seeds will fall on bare rock surfaces, others in bodies of water, and many others in spots that are already effectively pre-empted by other plants. Few if any will reach an unpreempted soil surface. Here they may lie for months or years until rains have softened the seed coats. Then, provided the temperature is not too low, the embryos will be awakened by the moisture that diffuses in to them. Growth will be resumed at the point where it was arrested long before; the embryo will push a first root through the softened seed coat, and shortly the young sporophyte will be completely established in the soil and the sunlight.

We have stated that Gymnosperms are usually evergreens. This feature, however, will not distinguish the group sharply from Angiosperms; for actually there are a few Gymnosperms that are "deciduous" (i.e., shed their leaves every winter), and a few Angiosperms that are evergreen. We have stated that Gymnosperms are cone-bearers. This feature, too, fails to be quite distinctive, for there are some of the more primitive Angiosperms which bear reproductive structures that are virtually cones or strobili rather than true flowers. The real distinction between the two groups is referred to by their titles. Gymnosperm means "naked seed." The seeds, as we have seen, are produced on the exposed surface of the megasporophyll. Angiosperm, on the other hand, means "encased seed," for in the latter group, as we shall see, the seeds are produced within an inclosing case of tissue.

CHAPTER XVII

THE LIFE-CYCLE OF THE FLOWERING PLANT

THE culminating attainments in the evolution of the plant kingdom appear among the Angiosperms. Although there is tremendous variation as to body form in this vast assemblage of over 130,000 species, there is surprising uniformity with respect to the main features of the reproductive program. This uniformity suggests that the reproductive features involved are of great biological value, that they have been at least somethat responsible for the obvious success of the great group in which they occur.

In general, the Angiosperms are flowering plants. This feature, however, does nor provide a sharp or true distinction between this group and the cone-bearing Gymnosperms; for actually there are quite a number of the relatively more primitive members of the Angiosperm group which bear reproductive structures that are more like cones than flowers. The true distinguishing feature for the Angiosperms is described in the name itself—Angiosperm means "incased seed." As we shall see, all Angiosperms produce their seeds within an inclosing case. This is true of none of the Gymnosperms, which uniformly produce "naked seeds."

The Angiosperm group is sufficiently large to show great variation with respect to such relatively unimportant features of the flower as the number, size, and shape of the various floral parts. Without pausing to concern ourselves with this variation, we propose to describe the life-cycle of the Angiosperm in terms of an "ideal" flower, i.e., one which stands for a rough "average" condition for flowers in general (Fig. 105).

In such an average or typical flower we would find four sets of floral parts, arranged concentrically. The outermost cycle is composed of parts known individually as "sepals" (and sometimes referred to collectively as the "calyx"). These are usually leaf-like in both color and structure, though commonly smaller in size than the purely vegetative leaves that are born elsewhere on the sporophyte body. In its early developmental stages the flower is arranged in the form of a "bud," with the sepals folded together on the outside, inclosing and protecting the more delicate structures within. Here, then, we encounter one more example of tissues which incase and

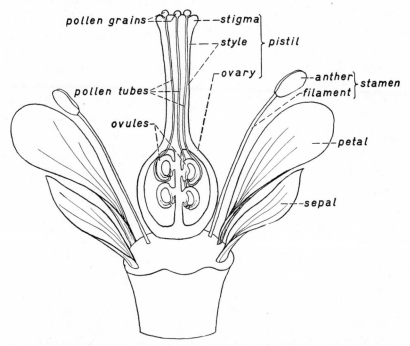

FIG. 105.—Diagram of an ideal flower, shown in longitudinal section. (The pollen grains and pollen tubes are disproportionately large, as is the free space shown within the ovary cavity.)

protect certain critical structures of the organism while they are passing through the tender stages of early development, and later unfold and expose those structures when they are mature and prepared to function in the medium of air.

With the sepals unfolded, we would encounter just within them a cycle of "petals" (sometimes referred to collectively as the "corolla"). These are commonly larger than the sepals (when the flower is mature), more delicate in texture, and endowed with a bright coloration. To the majority of people the flower is nothing more than an object of aesthetic value, and this value is contained largely in the petals, which, in most cases, are by far the most conspicuous part of the flower. But nature had evolved the petals of flowers long before she had evolved human beings, and their evolution was directed into channels of value to the plants that produced them. The value of petals lies in their power of attracting insects, which, in the majority of Angiosperms, are necessary agents for the completion of the life-cycle.

Inside the petals there comes a cycle of "stamens," each one consisting of a long, thin "filament," surmounted by a club-shaped "anther." The appearance of the stamen is suggestive of a spore-case on a stalk; and the appearance is not deceptive, for the anther of the stamen is indeed a spore-case, containing numerous tiny spores, and the filament is a stalk which so elevates the anther as to put the spores in a favorable position for distribution when the proper time comes.

At the center of the flower lies the "pistil."* This is the most critical part of the flower, for it is here that the seed (or seeds) will be produced. Three parts are usually recognizable in the pistil: at the bottom is a swollen "ovary"; extending upward from the middle of the ovary a stalk-like "style"; and at the top of the style a small platform of tissue known as the "stigma." As we shall see, each of these three parts has a distinct function in connection with those events which lead to fertilization and the production of the seed.

In earlier chapters we saw how the life-cycles of moss and fern pursued a regular and a similar course in which the following stages were encountered successively: zygote, sporophyte, spore, gametophyte, gametes, zygote. Every complete generation for the organ-

---

* For the benefit of those who have read chap. xvi, the term "pistil" bears the following relationship to the term carpel, or megasporophyll. Among the numerous species of Angiosperms, the number of carpels in a flower varies from one to many. When there are several carpels they may be structurally distinct, or they may all be united into a common structure in the center of the flower. "Pistil" refers to the structurally distinct unit, whether that be a single carpel or a composite of several.

ism contains a great many generations for the individual cells, for it is a long series of cell divisions that carries the organism from zygote around to zygote again. One might think of a toy train moving steadily around its circular track, save for just one stage which this analogy would fail to depict. Throughout most of the cycle the heritage of the species is moving along a single track, but at the stage of the gametes it is moving for a very short time along double tracks, which meet again in the zygote. In the case of the seed plant, double tracks occupy a greater portion of the entire cycle, for the separation

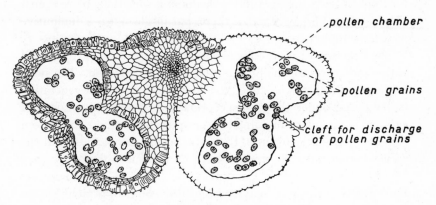

FIG. 106.—Cross-section of a mature anther

occurs at stage of the spore and continues until the zygote is reached. So we may expect to encounter two types of spores, followed by two types of gametophytes and two types of gametes.

For our purposes it will be sufficient to distinguish the spores as "small spores" and "large spores."* Let us first give our attention to the small spores. When the stamen is mature, the anther at its upper end contains two chambers full of these small spores, or "pollen grains," as they are commonly called (Fig. 106). At first each spore is organized as a single cell; within each spore coat there is just the single nucleus surrounded by cytoplasm. Even before the time of pollen shedding, however, the contents of each spore have divided

---

* These, of course, correspond to the "microspores" and "megaspores" that were discussed in chaps. xii and xvi.

up into three diminutive cells.* This three-celled body, contained entirely within the spore wall, is actually a gametophyte. Far smaller than any other gametophyte in the plant kingdom, it produces no recognizable sex organs. Instead, two of its three cells become organized as sperms. Hence we know it to be a "male gametophyte," since it produces gametes of only this one sex. It is noteworthy that the sperms are not ciliated. At last, as we shall see, the plant kingdom has evolved a method of getting the sperms to the eggs which does not depend upon the presence of a film of water. Note also that the "pollen grain" at the time of shedding is something more than a simple spore. The spore wall is there, but inside of it is a diminutive male gametophyte.

The large spores are produced within the base of the pistil, buried within structures to which we must give some attention. The swollen base of the pistil was called the "ovary" by an earlier generation of botanists, under the misconception that it was a container of eggs, as is the ovary of animals. As we shall see, there are indeed eggs deep within the ovary of the pistil, but it was not these that the earlier botanists had in mind. When they cut into the ovary they found within a cavity that was almost filled with small, ovoid objects that grew into the ovary cavity from their points of attachment on the ovary wall or axis. These objects bore some superficial resemblance to animal eggs. So the older botanists called them "little eggs" or "ovules," and it was for this reason that their container was called the ovary. Now the ovule of any Spermatophyte is an exceedingly significant structure, for it is the structure which, following certain transformations, is to become the seed. But the ovule is far from being what it was first thought to be, a "little egg." Instead it is actually a spore-case, but one which differs in several important respects from the spore-cases that we have encountered hitherto.

The wall of the spore-case of Thallophytes was nothing but a cell wall, within which an internal division of the protoplasm occurred to produce the spores. The spore-case of the fern was a many-celled structure, but its wall consisted of no more than a single layer of epi-

---

* The statement above is a deliberate simplification. Pollen grains at the time of shedding commonly contain only two cells. It is at a later time that one of these two divides to bring the total number up to three cells.

dermal cells. In the ovule of the seed plant, however, we have a spore-case that is comparatively thick-walled, the wall being composed of several layers of sterile cells (Fig. 107).*

All previous spore-cases contained numerous spores. The ovule of the seed plant is peculiar in containing just one.† Buried deep in the center of the tissues of the young ovule is a single large cell which is to act as the spore. This is the "large spore" of the life-cycle. The young pollen grains, as we have seen, were the "small spores."

But the most significant of all the peculiarities of the ovule lies in its failure to shed the spore. All spore-cases that we have seen to date have broken and released the contained spores at the time that they were ripe for shedding. The ovule, however, retains its single spore indefinitely, and it is this feature that makes it possible for the ovule and its contents to become the seed.

We have seen how the small spore produced tiny male gametophytes within the confines of the spore wall. The large spore within the ovule likewise produces a gametophyte—in this case a female

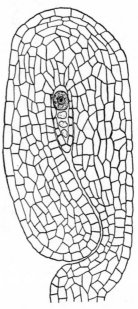

Fig. 107.—Longitudinal section of a young ovule (of *Erigeron*). Four potential spores have been formed, but three are being aborted and only one is developing further.

gametophyte. As compared with the gametophytes of lower plant groups, the female gametophyte of Angiosperms is markedly reduced

---

* Actually we have encountered what have appeared to be thick-walled spore-cases on two previous occasions. The capsule of the moss and the anther of Angiosperms themselves may both be thought of as complex spore-cases. In each of these situations the spores were contained within walls that were several cell layers in thickness. It is the great functional significance of the wall of the ovule that leads us to make a special point of its thickness.

† Actually four potential spores are produced. Three of these, however, are aborted, leaving only the one functional spore.

in size. As compared with the male gametophyte of Angiosperms themselves, it is large, corresponding to the greater size of the spore which contains it. At first the large spore contains only the one

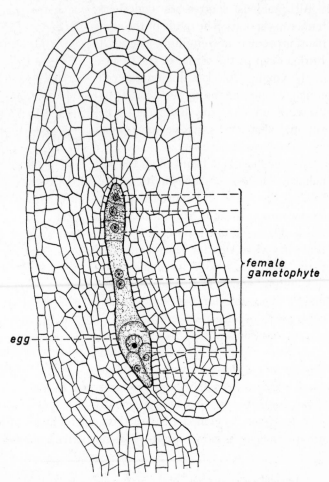

FIG. 108.—Longitudinal section of an older ovule (of *Erigeron*), containing a mature female gametophyte.

nucleus, together with a large supply of cytoplasm. In its subsequent development the spore passes through a series of three nuclear divisions, so that the mature female gametophyte contains a total of eight nuclei. We can disregard seven of these; the eighth, which usually becomes larger than the rest, takes up a position near one end of the gametophyte. A membrane surrounds this nucleus together with a certain amount of cytoplasm, and the resulting cell constitutes the egg. Once again we find no recognizable sex organ, but a direct production of the gamete by the gametophyte. Whereas the tiny male gametophyte produced two sperms, the decidedly larger female gametophyte has a total output of only one egg (Fig. 108).

The stage is now set for fertilization, but the bringing of the sperm to the egg is a much more complicated process in the Angiosperm than in the lower plants that we have examined. There are two distinct chapters or stages in the process, the first being "pollination," and the second being the growth of the "pollen tube."

By "pollination" we refer to the transfer of pollen from the anther where it is produced to the stigma at the upper end of the pistil. In some species "self-pollination" is the rule, i.e., transfer of pollen to the stigma of the same flower. Here the problem is comparatively simple. By the action of gravity the pollen may merely fall from the anther to the stigma, which is carried at a slightly lower level in such cases. In other cases the anthers grow compactly together, forming a tube; when pollen is ripe the style elongates, forcing the stigma up through this tube, where it collects pollen by direct contact.

Self-pollination, as we shall see, inevitably results in self-fertilization, i.e., the union of egg and sperm that have been derived (indirectly) from the same sporophyte individual. A sustained program of self-fertilization tends to perpetuate the species unchanged. It does not absolutely preclude the possibility of evolutionary change, but makes for a slower evolutionary progress than might result if occasional cross-fertilization occurred.* It is not surprising, therefore, to find that the majority of Angiosperms exhibit at least a certain amount of cross-fertilization. This is accomplished, of course, by devices which favor, and in some cases absolutely insure, cross-pollination.

---

* In species that are self-fertilized, variation may be introduced on rare occasions by the phenomenon of "mutation" (see chap. xix). This makes evolution possible, but since the variations introduced are few the evolution which results is comparatively slow.

In cases of cross-pollination, pollen transfer presents a real problem. The details of solution of this problem are so highly variable among the numerous species of Angiosperms that we cannot here hope to cover the question. Suffice it for our purposes that, in general, two great agencies are employed for the transfer of pollen from the flowers of one plant to those of another. These two agencies are wind and insects.

With certain exceptions, wind-pollination seems to be the more primitive of the two methods. Here we encounter another example of the remarkable timing devices that have been worked out in the organic world. During the comparatively brief period when pollen is being shed by one plant, the stigma surfaces of other plants in the vicinity are exuding a sticky liquid which equips them to catch and to hold such pollen grains as may happen to alight there. Though this feature may impress us as to the "efficiency" of nature, there is an accompanying feature which leaves the opposite impression. Pollen grains are light, and are so equipped with walls as to withstand the drying influence of the air for a certain time, but they have no equipment whatever for active movement, and there is nothing to direct them to their proper destination. They are absolutely at the mercy of chance currents of air, which may carry them anywhere in a field in which the stigmas of flowers of the same species present targets that are both tiny and few. The plant meets this problem by producing a tremendous quantity of pollen, the vast bulk of which comes to naught. This may seem wasteful, but it is not a meaningless, prodigal wastefulness, for it is necessary to insure a reasonable amount of effective pollination under this obviously inefficient program of wind-pollination.

Two incidental points are worth noting. Wind-pollination charges the air with a profusion of microscopic pollen grains. Reaching the eyes and the noses of certain people, these produce the well-known effect of "hay fever," an example of the non-beneficent interaction of different organisms. The other point to be noted is that wind-pollinated flowers have petals that are usually small and not brightly colored, or may be lacking completely; for the petal has no adaptive value in such plants.

The other main agency for pollination is insects. Here we have a much more efficient device, for it moves pollen with surprising directness and accuracy from source to destination, and the plant can solve its pollination problem with a total output of pollen that is far less than in the case of the wind-pollinated species. Insect-pollination is the most conspicuous and the best known of the many existing examples of interaction between living organisms of different types to the advantage of each. The value of the insect to the flower is efficient cross-pollination (sometimes self-pollination). The value of the flower to the insect is a food value. Usually the food is in the form of "nectar," a sugar solution secreted by certain glandular cells in the flower, though in some cases the insects devour a part of the pollen supply itself. The brightly colored petals and the odors which some flowers exude have an important indirect value in advertising the presence of the flowers to the insects.

Here we encounter an even more striking example of "timing" in nature. Within the same brief period pollen is ripe for shedding, petals are at their largest and brightest, odors their most intense, nectar being secreted, and the stigmas of other flowers in the proper condition for reception of pollen. In the process of pollination the insect thrusts all or part of its hairy body down into the tube of the flower in quest of the nectar, which is produced by certain cells at the bottom. The physical characteristics of flower and insect are such that this movement inevitably rubs portions of the insect body past the shedding anthers. On its visit to the next flower these pollen-smeared portions of the insect brush past the sticky stigma, and thus the pollen arrives at its proper destination.

The foregoing statement reveals the main principles involved in insect-pollination, but fails miserably to do justice to the tremendous diversity of fine adjustments to this process that have been evolved in the bodies of the many pollinating insects and in the many flowers which they pollinate. Not only is the size, shape, and general physical equipment of a particular insect remarkably adapted to the physical peculiarities of the (one or a few) types of flowers that it pollinates, but its instincts direct it in making exactly the appropriate movements. The fossil record tells us that primitive insects were first evolved about two hundred and fifty million years ago,* and that the number of insect types was increasing rather slowly in the

---

* Toward the end of the Paleozoic era.

following time periods. About a hundred and twenty million years ago,* however, the evolution and diversification of insect types took on new impetus. It has proceeded at such a great rate ever since that today the insect species far outnumber the total of all other groups in the animal kingdom. The new impetus to the evolution of insects was the appearance of flowering plants, and during this period when insects were evolving into myriad types Angiosperms were doing the same, so that today they, too, outnumber all other groups within their own kingdom. The adaptive interrelationship of insects and flowers has been so perfected that the great Charles Darwin, about 1870, when presented with a new flower of rare dimensions, was able to predict correctly that a moth of corresponding dimensions would be discovered living in the same region.

Once that pollen grains have arrived upon the sticky surface of the stigma they behave in a manner reminiscent of the spores of parasitic fungi. The protoplasm within expands and pushes through the spore walls like the young mycelium of a fungus. The tiny filament that is extruded is known as the "pollen tube." Very rapidly it burrows its way down through the stigmatic surface and into the tissues of the style. This pollen tube is the male gametophyte which has escaped from the confines of the spore wall and has related itself to the tissues of the pistil very much as a fungus mycelium relates itself to the tissues of its host. Rapidly it progresses through the process of eating its way, until it has completely traversed the style and has arrived at the ovary cavity. Directive forces that are little understood guide the growing pollen tube to the nearest ovule. Penetrating the ovule tissues, the growing tube comes at last to lie in actual contact with the female gametophyte. Meanwhile other pollen tubes, following much the same general course, are directed to pass farther and penetrate other ovules which have not been pre-empted (Fig. 105).

Before the pollen tube has emerged very far from its confinement within the old spore wall, the two male gametes slip downward to take a position near the tip of the advancing tube. The sperms are not ciliated—Angiosperms have no need of that—but are moved to their new position by streaming movements of the cytoplasm within

---

* During the latter part of the Mesozoic era.

the male gametophyte. Hence when the female gametophyte has
been reached there are two sperms in readiness at the tip of the tube
(Fig. 109). At this stage a rupture in the end of the tube is followed
by extrusion of the two sperms, so that they are now free within the

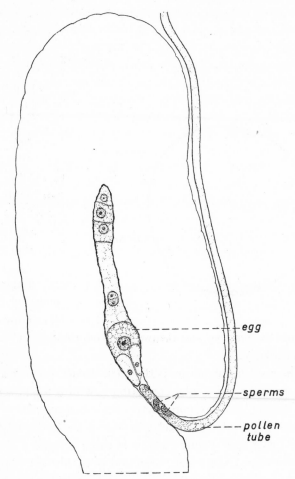

FIG. 109.—Pollen tube entering the mature ovule (diagrammatic)

confines of the female gametophyte. One of the two sperms fertilizes
the egg, and thus at last the zygote is produced.

As in Bryophytes and Pteridophytes, the Spermatophyte zygote
immediately gets under way with the production of the young sporo-

phyte. Very rapidly this developing embryo obliterates the old female gametophyte and comes to occupy the corresponding space within the ovule. Then rather suddenly the embryo stops growing and passes into a state of dormancy.

In earlier chapters we noted that living tissues were often thrown into a state of dormancy by a cutting-off of their water supply. This is exactly what happens to the embryo sporophyte. In some manner that has not been analyzed, either fertilization itself, or some influence of the pollen tube, provides a stimulus which initiates changes in the wall of the ovule. These wall tissues which have hitherto been soft, start to undergo a transformation, which is occurring at the same time that the embryo sporophyte is developing within. Soft tissues are gradually transformed into a tough (and in some cases an exceedingly hard and thick) wall, which at last completely surrounds the embryo and throws it into dormancy by cutting off its water supply.

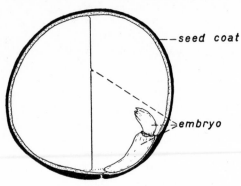

FIG. 110.—Section through a bean seed

The resulting structure is the seed. Hence a seed may be defined as a transformed ovule, consisting now of a tough seed coat surrounding an embryo sporophyte. These two components are always present in a seed, and sometimes there is also a third. For its future development the young embryo is going to need a good supply of food. In some seeds (e.g., bean and pea) this food supply is stored within certain parts of the embryo itself. Hence when we peel off the coat from a seed of this sort we encounter nothing but a rather large embryo within (Fig. 110). In other seeds (e.g., corn and wheat) the food is stored not in, but around, the embryo. Within such a seed one finds a rather small embryo imbedded in a larger food storage tissue ("endosperm") (Fig. 111).* In either event, the food thus stored for the future nourishment of the embryo is often exploited by man and devoted to his own nourishment.

The release and distribution of seeds and the emergence of the young sporophyte are matters that we shall discuss in the next chapter. For the present, however, let us consider the biological value of the seed. Biological value it must have, for the seed is the structure that is largely responsible for the overwhelming success of the Spermatophytes.

In an earlier chapter we called attention to a great evolutionary trend which appears in both animal and plant kingdoms. Among the more progressive lines of descent we find that both animals and plants have increasingly made provision for their young; so that the highest animals of today and the highest plants of today have perfected devices for giving their offspring a good start in life. The seed is such a device. The food stored in or around the embryo sporophyte will enable it, when the proper moment arrives, to make rapid growth and quickly to establish itself as a self-supporting plant. Even

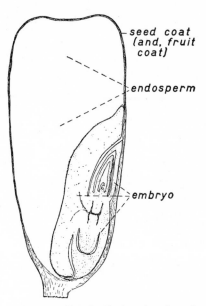

FIG. 111.—Longitudinal section through a corn grain. (The so-called "grain" is technically a "fruit" rather than merely a seed, for the coat of the seed proper is closely surrounded by a thin layer of tissue that was derived not from the ovule itself but from the ovary wall.)

---

* This endosperm tissue originates in a surprising manner. When one sperm from the pollen tube fuses with the egg to form the zygote, the second sperm moves a little farther and fuses with two other nuclei of the female gametophyte. As far as the writer knows, this program is followed faithfully among all Angiosperms. It is spoken of as "double fertilization" and brings into existence a "triple fusion nucleus" or "endosperm nucleus," which, in its heritage, is a

more significant, however, is the fact that the earliest and most sensitive stages of sporophyte development have occurred in a well-protected place, attached to the body of the parent. A third biological value of the seed is its value as an agent of distribution, and a fourth lies in its power to tide the species over a period of hard conditions; these are matters that will be treated in the following chapter.

---

strange sister to the zygote. At the same time that the embryo sporophyte is developing from the zygote, an endosperm tissue is developing from the endosperm nucleus, and it is the encroachments of both which obliterate the old female gametophyte. In some seeds (e.g., bean and pea) the voracious embryo now turns upon the young endosperm tissue and devours it completely; so that the mature seed contains nothing but embryo within the seed coat. In other seeds (e.g., corn and wheat) the endosperm not only persists, but comes to occupy a larger part of the mature seed than does its sister, the embryo. Even here, however, the endosperm is destined later to be consumed by the embryo, so that the sporophyte of every Angiosperm is guilty of cannibalism at some stage.

CHAPTER XVIII

SEED DISTRIBUTION AND SEED GERMINATION

NEAR the end of the last chapter we left the seed, along with its companion seeds, inclosed within the ovary of the pistil. To accomplish any useful purpose these seeds must emerge and get distributed. How seeds emerge and how they may be distributed depend, in large part, upon the surrounding structures. If we examine these surrounding structures we find that they, too, have undergone transformations. While the seeds have been ripening, the thing that we call the "fruit" has been ripening around them.

When the layman uses the term "fruit" he is referring to edibles of a category that would be rather difficult to define. When the botanist uses the term "fruit" he is referring to the part of the plant in which the seeds are contained. In most cases this structure is nothing but the ripened ovary. An orange, for example, is an ovary that has ripened into a fleshy, juicy condition. This would be recognized as a fruit by both the botanist and the layman. Botanically speaking, however, the tomato is likewise a "fleshy fruit," though the layman would be likely to exclude this from his classification on no better grounds than that it was grown in a garden instead of an orchard (Fig. 112). Botanically, a nut is also a fruit, and the pod of a pea, for these are also ripened ovaries containing seeds; though "dry fruits" of this sort are never recognized as fruits by the layman (Fig. 113). In some cases the fruit is more than a single ovary. A raspberry, for example, is an aggregate fruit, in which each tiny lobe has been derived from the ovary of a separate pistil; and a fruit like the pineapple is even more elaborate, being the fused product of a great many flowers, and including not only the transformed ovaries but other floral parts as well (Fig. 114).

It should be realized that the fruit that we buy at a fruit-stand is usually a very different thing from the fruit that was evolved by Angiosperm plants growing in a state of nature. During the past few centuries man has directed the course of evolution along lines which catered to his own interests. Starting with the relatively puny fruits which nature provided, he has gradually led his cultivated

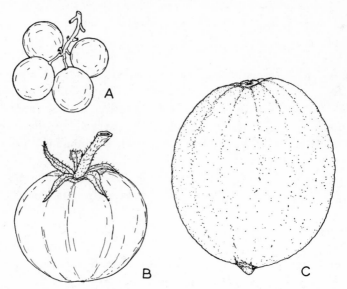

FIG. 112.—Simple fleshy fruits: *A*, grape; *B*, tomato; *C*, orange. In these cases the entire fruit is derived from the tissues which composed the wall of a single ovary.

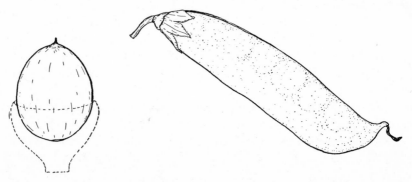

FIG. 113.—Dry fruits. On the left, a "nut" (the acorn of the oak in this case). On the right, the "pod" of the pea.

plants to produce bigger, tenderer, juicier, and tastier fruits. We moderns, with our appetites adjusted to the charms of modern cultivated fruits, would in most cases turn up our noses if offered the fruit of the original, wild, ancestral types.

But nature did not produce the original fruits merely in order to

Fig. 114.—Complex fleshy fruits. On the left, the raspberry, in which each lobe has been derived from the ovary of a separate pistil, but all the pistils involved have been parts of the one flower. On the right, the pineapple, produced by the transformation and fusion of the tissues of a great many distinct flowers.

nourish and delight mankind. Fruits were in existence for millions of years before man put in his appearance, and during all this time they must have been serving a useful purpose for the species which produced them. The ripe fruit has one minor biological value in carrying on the function which was started by the young ovary, i.e., protecting the seeds until they are ripe to be released. In the case of some fruits, there may be some biological value in "manuring" the little spot of ground in which the young sporophyte is to take root; for a fallen fruit will disintegrate and enrich the soil at the very place where the contained seeds will sprout into young plants. But undoubtedly the main biological value of the fruit is in providing just one of the many means of accomplishing seed distribution, as we shall see shortly.

Among the three lower divisions of the plant kingdom, the spore is prevailingly the agent of distribution for the species, and a rather effective one, too, owing to its small size and the ease with which it may be carried through air or water. With the revision of the lifecycle that appears among seed-plants, however, the spore can no longer play this rôle. The "large spores" are never released, as such, and the "small spores" or pollen grains—if they are to function at all —must do their functioning at spots where the ovules are already located. With spores eliminated as a means of distribution, and the plant body itself unable to migrate, there remains only the seed as an effective agent for spreading the species over an increasing territory. The widespread present distribution of seed plants is enough to tell us that seeds must have played this rôle very effectively during the past.

Seed distribution is accomplished by many different devices. To a very limited extent, gravity may effect distribution, as when nuts fall to the ground and roll down a slope. Much more important is water, as in the case of those seed plants which grow actually in the water, as well as those which we commonly encounter on the banks of streams and rivers. A few plants accomplish seed distribution on a small scale by devices of the fruit which make for a forceful ejection of seeds, as in the "touch-me-not" and the "squirting cucumber." But the major agencies for seed distribution are the animals and the wind.

Animals appropriate both dry and fleshy fruits as articles of diet. If animals were to consume the entire contents of every fruit which they picked off the plant, the net result would be adverse to the success of the plant species. Such, of course, is not the case, or fruits which so attract the animals would probably not have been evolved. A few animals, such as the squirrel, often actually plant seeds, by burying nuts and failing to return for them. There are many more that eat the fruit and eschew the seeds, and many that eat the seeds at the same time but pass them unharmed through their alimentary tracts.

Edibility is not the only device for winning the co-operation of animals. Some plants form burrs through barbed extensions of the seed coats or coats of the dry fruits which surround the seeds (Fig. 115). Furred animals may carry these burrs for great distances before dropping them. The small seeds of a good many plants, coming to rest on the muddy banks of streams and ponds, may be carried tremendous distances through adhering to the feet of migratory birds. And we must not forget that, among the animals, man himself, voluntarily and involuntarily, has effected a great deal of seed distribution in the course of his migrations.

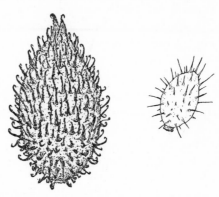

FIG. 115.—Burrs: cockleburr on the left; grass burr on the right

Wind is employed even more extensively as an agent of seed distribution. Some seeds (e.g., grasses) are tiny, flat, and light, and can readily be borne on the wind. Others, somewhat larger and heavier, make use of the glider or parachute principle. In the pine, and in many other Gymnosperms, an extension of the seed coat itself forms a passive wing, while in maple, ash, and elm, wings are formed by extensions of the dry fruit. In milkweed, in the cottonwood, and in the cotton plant itself, the seed coat is covered with many fine hair-like outgrowths which provide such great air resistance as to keep the seeds aloft for a long time through the action of chance air currents. The dandelion fruit is provided with a beautiful little parachute which has been derived from certain outer regions of the flower (Fig. 116). In the tumble-weed, the entire plant goes bouncing along in the wind, dropping seeds as it goes.

The period of dormancy of the embryo is quite variable. For plants that live in the temperate zone the commonest program is one in which the embryo lies dormant for about half a year, since most seeds are produced in late summer and early autumn, lie on the ground throughout the ensuing winter, and sprout forth with the new plant-lets during the following spring. Some seeds, however, demand a longer rest period than this, and will not "germinate" before the sec-

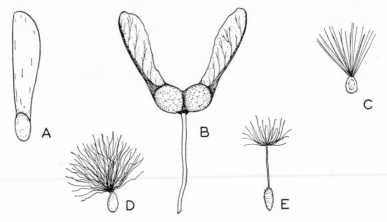

Fig. 116.—Wind-borne seeds and fruits: *A*, pine seed; *B*, maple fruits; *C*, milkweed seed; *D*, cottonwood seed; *E*, dandelion fruit.

ond season, or even later. Even those seeds that normally germinate the following spring will retain their viability for several years if kept under dry and relatively cool conditions. The old stories about the germination of seeds that were found buried with the Egyptian mummies have been entirely discredited, but it is known that some seeds will last for over fifty years, and there is one creditable report on the lotus seed that points to a period of at least one hundred and fifty years.

The expression, "seed germination" refers to all that happens in connection with the emergence of the young sporophyte from within the seed and its establishment as an independent plantlet. Primitive agricultural races have always been greatly impressed with this phenomenon and have often dedicated to it some form of religious ceremony. For the coming into existence of "new life" was always .impressive, and to the untrained the germination of seeds in the spring was a large-scale example of this miracle. We realize today that strictly new life is never* brought into existence, but that new organisms are derived only by a process of reproduction from parent-organisms of the same general type. We realize further that seed germination does not even mark the true beginning of a plant genera-tion, as does the zygote or the spore. Seed germination is merely the awakening of a young plant that has been dormant for a time; it now has the opportunity to carry on from the point at which its develop-ment had been arrested.

Development had been arrested when and because the water sup-ply had been cut off by the completion of the seed coat. Develop-ment recommences when the water that is provided by spring rains and melting snow soaks through the softening seed coat and reaches the embryo. Though water might be regarded as the most important factor, temperature is also involved, for germination will never be a success if the surrounding temperature is significantly below (or above) the temperature range to which the species is adjusted. Oxy-gen, too, is a vital necessity, for respiration goes on at a great rate in the newly awakened embryo.

The fuel for respiration and the materials for growth are derived from the food that has been stored in the seed, either within the em-bryo itself or surrounding the embryo. Growth does not occur simul-taneously at the same rate in all parts of the embryo. While other parts are still growing quite slowly, if at all, that part of the embryo which is to become the first root starts to develop at a great rate. This prospective root is located just within that region of the seed coat which is thinnest and which first becomes well softened by the water. Here the fast-growing rootlet pushes through and for quite a while is the only part of the embryo to have emerged. Positively geotropic and hydrotropic, the young root curves (if necessary) to

---

* At least never within the period and scope of man's critical observation. At some time in the past, and perhaps under conditions that no longer exist on earth, the first living organisms were probably derived from non-living ante-cedents.

thrust itself into the soil. Rapidly it establishes itself and commences an intake of water and soil salts, which are moved upward through the simple beginnings of the vascular system to the parts of the embryo which are still within the seed.

Even a very young embryo is organized on a distinct axis, with the young root tip and stem tip at opposite ends. As we have seen, the first chapter in growth is devoted to the extension of the root.

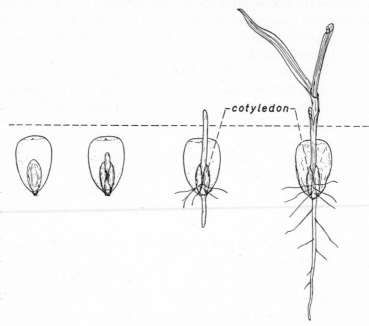

FIG. 117.—Successive stages in the germination of a grain of corn

After the root has become established, a second chapter commences, in which other portions of the embryo initiate an active growth, while the growth of the root continues. In a manner which differs in detail among seed plants, the growth of these higher embryo regions serves to extricate the young stem tip, either by pushing it straight through the seed coat or by pulling it out backwards. As with the root, the young stem, when it first emerges, may not be in the proper orientation. Very soon, however, its negative geotropism and its positive phototropism will direct it upward into the sunlight. Even the embryo within the seed had possessed the tiny beginnings of the first leaves near the upper end of its axis. With the emergence of the stem these leaves develop rapidly; with the exposure to sunlight,

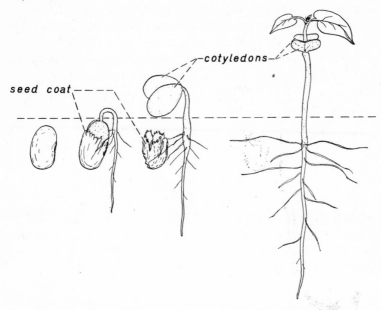

FIG. 118.—Successive stages in the germination of a bean seed

chlorophyll rapidly appears; and in a surprisingly short time the young sporophyte is fully established as an independent plant (Figs. 117, 118).*

---

* This much simplified account of seed germination has ruthlessly glossed over quite a number of significant details.

In the classification of Angiosperms, the first subdivision to be made is into two groups which we call Dicotyledons and Monocotyledons. These two groups differ in a number of characteristics (see chap. xx), but the difference from which the two groups derive their names is a difference in their embryos. The primary organization of the embryo is the same in each group, with root tip and stem tip marking the two ends of the axis. In addition the embryo possesses certain members which grow laterally from this major axis. These structures are known as the "cotyledons," and the names, Monocotyledon and Dicotyledon, bespeak the possession of one and of two cotyledons, respectively (Fig. 119). There are some species in which cotyledons emerge from the seed coat along with the stem, produce chlorophyll, and actually conduct a certain amount of photosynthesis, later withering away after the first true leaves are well developed. This, however, is more the exception than the rule, and, in general, cotyledons have merely one of the following two functions. Among most Dicotyledons, where the embryo proper fills the entire seed, the two cotyledons are the regions of food storage. In such cases, as is readily illustrated by a

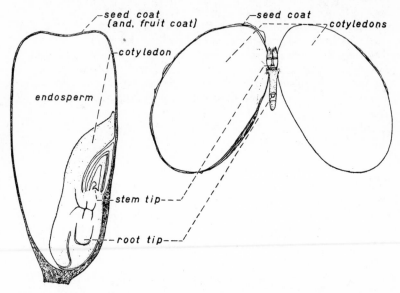

FIG. 119.—Embryo types. On the left, a longitudinal section of the corn grain, revealing the monocotyledonous embryo. On the right, the dicotyledonous embryo of the bean. In the latter the two cotyledons, which are actually folded closely together within the seed, have been artificially separated to show the relationships, while the main axis of the embryo is shown in partial section to reveal the true stem tip and root tip, which are actually surrounded by other embryonic tissues at this stage.

bean or a pea, the cotyledons constitute by far the greatest bulk of the embryo. Among most Monocotyledons, where the embryo is comparatively small, and surrounded by a food storage region, the single cotyledon acts somewhat as does the "placenta" in the higher mammals, forming into a flat pad which is appressed to one part of the food storage region, and through which food is absorbed for the benefit of the other parts of the embryo.

Whether food is stored within or without the embryo, the storage form is usually starch. Starch is a compact and efficient form in which to store up a great deal of energy; it is virtually insoluble and must be transformed to some soluble form if it is to be moved through a series of cell walls to the cells which are in need of food. The situation demands the process of "digestion," in which starch is broken down to a simple, soluble sugar. Rapid digestion, at the temperature range which is consistent with the life of animals and plants, is made possible by those remarkable biological substances which we term "enzymes." Though present in minute quantities itself, a special enzyme for starch digestion tremendously accelerates the digestion of starch into sugar, thus making possible its transfer and utilization. We see, therefore, that underlying and antecedent to the visible emergence of the embryo, there are certain physiological processes which are absolutely essential to seed germination.

As implied before, the mechanics of emergence of the young sporophyte are rather variable among seed plants. Some of the differences are illustrated by contrasting the programs characteristic of the corn and of the bean.

In the former, the monocotyledonous embryo lies just beneath the rather thin seed coat along one of the two broad, flat surfaces of the grain. First the young root and, shortly afterward, the young stem push directly through the thin surface covering at different points. This is illustrated by Figure 117, which does not, however, show the redirecting of stem and root that would occur, through the tropisms, if the grain happened to be planted in a different orientation.

In the bean, in which the large dicotyledonous embryo fills the seed, establishment of the root in the soil is followed by a rapid elongation in that part of the embryo axis which lies above the root, but below the point of attachment of the cotyledons. Since the root end of the axis is now firmly anchored in the soil, while the upper part of the axis, together with the cotyledons, is still trapped within the seed coat, the elongation of this middle zone of the axis inevitably results in its buckling into a characteristic "arch." This arch is like a tensed spring, and as development continues the tension increases, until at last the two cotyledons, with the stem tip between them, are forcibly pulled backward out of the old seed coat. Though the cotyledons may become green and conduct a little photosynthesis, their real usefulness is past, and they soon wither away. Meantime the first true leaves, which have been lying in embryonic form around the stem tip, enlarge and assume their proper function (Fig. 118).

CHAPTER XIX

THE PROCESS OF ORGANIC EVOLUTION

ANY survey of the plant kingdom must leave in the mind of the surveyor two very strong impressions: (1) He must be impressed by the almost innumerable examples of adaptation that are manifest in plant structures and plant functions. In case after case he finds the plant well equipped to maintain itself and to perpetuate the species in the environment in which it is growing. In very few cases does he find any conspicuous maladjustment. (2) He must also be impressed by the tremendous diversity of form that he finds among the quarter of a million species that make up the plant kingdom.

Before the nineteenth century most biologists accounted for these two phenomena by means of the "hypothesis of special creation." According to this hypothesis, each species of plant and animal originated through an independent act. Thus, at some time and place in the past, where there was no pre-existing organism of species A, the first representative of this species was suddenly brought into existence, and—the hypothesis goes on to assert—all subsequent members of species A have been derived by descent without modification from that first representative. It follows that all existing members of species A are not only like each other but like that original ancestor of long ago. As an independent event, which had no significant connection with this first act of special creation, there was suddenly brought into existence the first representative* of species B, and all subsequent members of that species are thought to have been derived without modification from this original ancestor. Similar assumptions were made as to the origin of all species. For the million-odd species of existing plants and animals, therefore, there must have

---

* The sudden creation of a *pair* of organisms was assumed to occur in the case of the higher animals and other types in which there is a sexual differentiation of individuals.

been just as many independent acts of special creation in the first place.

In terms of this hypothesis, the two phenomena referred to at the opening of the chapter receive the following interpretation. The adaptive features of organisms are attributed to the intelligence of the creator or creative agency. The diversity of plant and animal kingdoms is attributed to the large number of acts of special creation that occurred.

This "hypothesis of special creation" leaves many questions to be answered—questions as to the identity of the creative agency, as to the raw materials used, as to the methods by which they were originally synthesized and organized into the bodies of living plants and animals. The whole account has a distinctly miraculous flavor.

Those who today believe in special creation—and there are many —are basing their belief on "word of authority." They believe because some person or document in whom they have faith asserts that it is true. No one should be censured for basing his belief on word of authority when he can find no better basis for belief. And yet it has been the blind acceptance of word of authority that has apparently blocked human progress in the past more than has any other single agency. The lesson that modern science brings to the world is that it is far wiser to base one's beliefs upon the facts of nature.

Even so, most of the scientists of the eighteenth century accepted special creation as the true account of the origin of the many known species of plants and animals. Early in the nineteenth century the French biologist, Lamarck, attempted to swing biologists over to the alternative "theory of organic evolution." His attempts met with very little success, however, and it remained for the Englishman, Charles Darwin, to persuade the biological public to discard special creation and to accept organic evolution in its stead. It was in 1859 that Darwin published his famous *Origin of Species*, a book that has been said to have had a greater influence upon the thought of the Western world than has any other except the Christian Bible. The influence of this book caused biologists to swing over to an acceptance of organic evolution rather rapidly. The rest of the world has been following suit more slowly ever since.

The "theory of organic evolution" is based upon an unbiased, dispassionate observation of the facts of nature, followed by logical inferences drawn from those facts, and by a subsequent testing of the inferences by means of other facts. In other words, the "theory of organic evolution" is a product of the "scientific method."

According to the evolutionist, all existing species have been derived by the process of organic evolution, or gradual modification through descent, from one or a few primitive ancestral types. As we shall see, the adaptive features present in plants and animals and the diversification in type that appears in plant and animal kingdoms are both to be accounted for—not by the actions of an original creative agency—but by the inevitable play of certain forces of nature upon plant and animal populations throughout the hundreds of millions of years during which evolution has been going on.

The writer knows of no biologist today who does not believe in organic evolution. How could it be otherwise, when the study of nature yields many independent lines of evidence in support of organic evolution and not a single line of evidence that contradicts it? The structural and functional characteristics of plant and animal types, their anatomy, their chemistry, their behavior, when compared with those of other types, yield overwhelming evidence of relationship through common ancestry. The combination of perfections and imperfections in detail that characterizes any of the more complex organisms can receive a sensible interpretation only on the assumption that that organism has been derived from a remote ancestor with somewhat different characteristics. The history of changes that have been wrought in domesticated animals and cultivated plants during the last few centuries tells us that evolution can occur. The impressive history that nature has recorded for us in the fossil record not only tells us that evolution has occurred but points out for us many of the exact lines that evolution has followed. Organic evolution must remain a "theory," for one cannot hope to observe directly the things that occurred during the millions of years of the past, but it is a theory so well established as to constitute the very foundation of modern biology.*

---

* Above we have made only a very brief and sweeping reference to the various "lines of evidence for evolution," a topic which really deserves several chap-

Though biologists have been in agreement for some time on the point that evolution *has* occurred, there has been some disagreement as to just *how* it has occurred—as to just what interaction of forces, what machinery of nature, has resulted in this gradual modification through descent. Many explanations of evolution have been proposed, and of the many there are two which have exercised major influence upon biological thought.

In 1809 Lamarck not only asserted emphatically that organic evolution was true but went further to provide an explanation of how the process had occurred. Since the overwhelming majority of biologists were still staunch "Special Creationists," Lamarck's ideas were met by a storm of ridicule, and it was not until after his death that they received fair consideration. When fairly considered, they appeared so plausible as to win a rather widespread acceptance.

Lamarck based his explanation upon the "principle of the blacksmith's arm." He pointed out that, during the lifetime of the individual organism, any part of its body which is used a great deal will develop in size and functional capacity, while those parts that are not used will gradually deteriorate. Even today biologists agree that, in general, this is a perfectly valid principle. In other words, it appears to be quite true that living organisms may "acquire characters" as the result of special experiences in the lifetime of the individual. This is not in itself evolution, but is, according to Lamarck, a potential start from which evolutionary consequences may follow.

When young organisms are introduced into a new environment—a type of event that must have occurred repeatedly in the past—the tendency is for them to make certain adjustments in their mode of life, involving a greater use of certain of their parts, or perhaps a lesser. By the time these organisms are mature, therefore, they will have acquired certain characters, though the acquirements may involve only rather slight increases or decreases in the magnitude (or quality) of some of their structures (or functions).

---

ters. No student of biology should feel satisfied until he has done justice to this topic through other readings. Though the plant kingdom alone provides many evidences for evolution, the beginning student is usually more impressed by the evidence which is provided by animal material.

As a next step, Lamarck makes an assumption that is critical to the whole interpretation, and the assumption is so plausible that it was, and still is, accepted without question by a great many people. Any bodily change that occurs during the lifetime of the individual will, according to Lamarck, be passed on to the offspring. In other words, the "acquired characters will be inherited," so that the new characters will be manifest (to a certain degree) in the offspring *without any special experiences on their own part.* If now, as seems probable, the individuals of the second generation make adjustments to the same new environment in the same general manner as did their parents, the further acquirements of the offspring will be added unto those that were inherited. Hence an accumulation of gains (or losses) is made possible through this inheritance of acquired characters, and a population that remains for many generations in a novel environment will inevitably come to have well-developed, novel characters that are attuned to that environment.

It follows that the remarkable adaptation that we see in existing types is a product of racial experience that has been accumulated through the inheritance of acquired characters. The diversity that we see in plants and animal kingdoms is due to the fact that the various lines of descent have adjusted themselves to different environmental conditions.*

Plausible as this "Lamarckian doctrine" may appear, there are but few modern biologists that believe in it. On the basis of general observations alone, biologists had been inclined to agree that "inheritance of acquired characters" probably occurred. When the question was put to critical experimental test, however, most of them

---

* The adjustment to environment that occurs during the lifetime of the individual may or may not involve a conscious process. Of course only humans— and only the minority of humans—are conscious of evolution. In many other species of higher animals, however, a conscious element may enter into the process, for, according to Lamarck, it is the conscious effort of the animals to satisfy their needs that leads to the acquirements of adaptive characters. When it comes to the lower animals and the plants, the whole thing must occur on an unconscious basis. In these cases, Lamarck thinks of the various environmental forces as acting directly to mold the characters of the unconscious organism into new patterns.

felt obliged to change their views. The critical, controlled experiments that were directed at this question during the last of the nineteenth century and the first of the twentieth provided substantial evidence to the effect that inheritance of acquired characters does not occur. It is always difficult, of course, to prove that a thing is impossible, and perhaps one should not say that any such demonstration has been made in connection with the question under discussion. Indeed, there are still a few biologists that subscribe to the "Lamarckian doctrine." The majority, however, have definitely discarded it, for the overwhelming bulk of experimental work indicates quite clearly that offspring simply do not inherit the special acquirements of their parents.

With the inheritance of acquired characters ruled out, it would seem difficult to comprehend how the course of evolution could have been directed along those channels necessary to lead to the highly adaptive results that we see today. In 1859, however, Charles Darwin provided an answer to that question that seems even today to be essentially valid. Darwin's *Origin of Species* not only succeeded in doing what Lamarck had failed to do—i.e., in convincing the general biological public that organic evolution was true—but it went farther and showed how organic evolution could occur without the assistance of the inheritance of acquired characters.*

One of the several principles upon which Darwin builds his theory is that of the "geometric ratio of increase." All living species have a potential reproductive ratio of over one; i.e., if all reproductive units function and all offspring survive, the second generation will outnumber the first. Earlier we noted how even those organisms with a reproductive ratio of only two (bacteria and blue-green algae) tend to increase their populations tremendously under those conditions which favor survival, growth, and reproduction. At the other extreme, we noted how some of the fungi produced billions of spores. In all plants and animals there is this potentiality for a geometric

---

* During the latter part of his life, Darwin felt forced to admit that inheritance of acquired characters had played a minor rôle in evolution. It was the experimental work of a still later date that has inclined biologists to refuse to make even this small concession that Darwin himself made.

ratio of increase, a *tendency* to increase populations prodigiously in the course of a few generations.

As a result, any small population, entering a new locality, will rapidly increase in size until it has reached the "saturation point," beyond which limitations in the food supply (and other natural checks) will make any further increase impossible. Though the tendency to increase is still exerting itself, there is no actual increase beyond this point. In other words, the potential geometric ratio of increase and the natural checks have arrived at an equilibrium, which is thereafter maintained until such time as some outside agency may upset it. At any given time most species in most localities have already attained this equilibrium, so that the surviving population of the second generation is essentially equal to that of the first. The maintenance of this equilibrium necessitates that, of the thousands of individuals that start life each generation, only a few hundreds (or usually less) actually survive to maturity. The majority are eliminated during the early stages of life. In other words, in most natural situations "death is the rule and life is the exception."

Another of the principles upon which Darwin builds his theory is that of the "struggle for existence." He points out that, with rare and at best only temporary exceptions, life is essentially a struggle. Not only man, with his elaborate conscious plans, but also the simplest of animals and plants—which, so far as we know, are entirely unconscious—are engaged, at most times and at most places, in some sort of struggle to keep alive.

In part this struggle is the competition between organisms that inevitably results from the geometric ratio of increase. Since space and food supply are limited, animals compete with others of the same species to win food. Less obvious but just as vital is the competition that frequently occurs among plants of the same species. The available light, water, and soil salts are usually present in limited quantities. The more vigorous or more "fortunate" plants secure these essentials, while the others are "crowded out" and die. Competition is interspecific as well as intraspecific. Animals of several different species may be competing for the same food; and this sort of thing is even more likely to be true with the green plants, where all species are dependent upon material and energy in the same forms.

Different in form but no less rigorous is the competition that occurs between the dependent organisms and those other organisms that provide their natural food supply. As we have seen, green plants are the victims of bacteria, fungi, and animals. It follows that the life or death of a given green plant will depend very largely upon the activities of these other organisms. Similarly, carnivorous animals are secondarily dependent upon herbivorous animals. Often there exists a "food-chain" of many links, starting with the green plant and ending with the dominant carnivore of the region. And wherever one link of this chain joins its neighbor, there is (with few exceptions) a life-and-death competition. In most natural situations these intricate interdependencies have been established, and an "equilibrium of species" has been approximated with about the same number of individuals of each species reaching the adult condition in each generation. But this state of equilibrium does not stand for a state of tranquillity. The equilibrium is not static but dynamic, involving a fairly continuous interplay of natural forces that are roughly balanced in their effects. Throughout this interplay of life, there is much death. Where life abounds, there also death abounds.

Competition with other organisms is not, however, the only thing that enters into the struggle for existence. In part, the struggle is imposed by non-living factors of the environment. Often animals struggle to get water, and plants to get water and soil salts, where no other organisms are at hand to compete with them, and where life depends upon the purely physical or chemical limitations of the environment and the ability of the organism to meet these limitations. Temperature often enters in to decide the fate of organisms; light also, and the physical or chemical nature of the medium or the substratum. With man, as with other organisms, the struggle for existence is usually quite apparent, and it is likewise quite apparent that it involves (1) competition with others of the same species, (2) competition with different species, and (3) problems of adjustment to the non-living forces of nature.

Another principle on which Darwin based his theory is that of the "universality of variation." We recognize without difficulty that no two human beings are exactly alike.* Were we as familiar with cats or with dogs or with any other species that exists, we would also acknowledge without hesitation that one could never find two individuals that would exactly match one another. With microscopic examination, we could establish the fact that no two blades of grass are ever quite alike, and, if our microscopes were a little better, we could doubtless establish the same thing for bacteria and other tiny organisms. So far as their *general* characteristics are concerned, all individuals of the same species are alike, but when it comes to the more *detailed* characteristics, a certain amount of variation is always present.

The biologist of today realizes that these variations are of two great types. Some of them are due to the fact that two individuals may have had different experiences. As a result of experience, therefore, or as a "response to environmental influence," the one individual may have one "acquired character" while the other individual has a different "acquired character" (or set of "acquired characters"). But even when two individuals of the same species have had the same experiences they may, and usually do, possess somewhat different characteristics. These latter expressed differences are due to differences in the hereditary endowments of the two individuals. In considering variation, therefore, we can distinguish "acquired differences" and "hereditary differences," and we *must* make this distinction if we are to see clearly how Darwin's theory accounts for evolutionary results.†

---

* Even "identical twins" can actually be distinguished upon sufficiently careful examination; and even if we *did* have to make an exception of such cases, the general validity of the universality of variation would be sufficient for the purposes of Darwin's theory.

† Though Darwin himself recognized the two types of variation, he very naturally failed to make the thoroughgoing distinction between them which was brought out so clearly by the experimental work of later biologists. Hence the modern biologist incorporates into his own explanation of evolution this refinement of Darwin's theory which was necessarily somewhat vague in Darwin's own mind.

Out of these several principles Darwin built an explanation of evolution that is usually referred to as "natural selection." Owing in part to the "ratio of increase," life is commonly characterized by a "struggle for existence." If all members of a species were identical, it would be purely a matter of chance which individuals won out in the struggle and succeeded in surviving. As it actually is, however, the struggling, competing population varies, and some of the elements of this variation may play a critical part in effecting the outcome of the struggle. Some, by virtue of their possession of a character which others lack (or by virtue of a higher grade of expression of that character than appears among the others), may thereby be somewhat better equipped for the struggle. The character in question may be a matter of size, strength, speed, rapidity of growth, efficiency of various protective devices, economy in the performance of certain functions, or any one of a great number of things. Darwin summed up all the possibilities by saying that certain organisms were more "fit" than were others, and by "fit" he meant better adapted to their environment. Darwin then simply pointed out the self-evident fact that the fit stand a better chance of surviving in the struggle for existence than do the unfit. Obviously, it is the survivors that are able to reproduce, while the non-survivors—that is, provided they perish before the reproductive period—will leave no offspring. Thus, the inevitable operation of natural forces leads to a certain amount of "natural selection," a selection of the more fit in the population to perpetuate the species.*

What, then, must be the character of the second generation if it

---

* Darwin's theory is often spoken of as the theory of the "survival of the fittest." The use of this phrase has unfortunately led some to jump to the conclusion that in every generation it is *only* the most fit that survive. Darwin himself cannot be held responsible for any such exaggeration of the facts. He was very careful to point out that some unfit usually survive and some fit usually perish as the result of chance. In other words, natural selection is not fully rigorous at every step. It is merely that there is a *differential* favoring somewhat the survival of the fit over the survival of the unfit. Obviously, the rigor of the selection will depend upon the severity of the competition and the severity of the non-living factors of the environment.

has been derived, in the main, from the more fit members of the first generation? In answering this question we must keep in mind the distinction between the two fundamental types of variation.

If an individual of the first generation possesses an "average" hereditary endowment, it may still, as the result of "fortunate" experiences, acquire characters which serve to make it fit. It may, therefore, be preserved by natural selection and leave offspring. Since acquired characters are not inherited, however, these offspring will be no better than were the average members of the first generation.

If an individual of the first generation has "average" experiences it may still be more fit than most, due to its superior hereditary endowment. Being more fit, it may be preserved and leave offspring. These offspring will indeed be better than were the average members of the first generation, for they will possess the superior hereditary endowments that were passed on to them by their immediate parents.

Nature, of course, in effecting the selection, bases her decision upon expressed characters alone. Hence the fit which she chooses are usually a mixed lot; in some the fitness is due to acquired characters, while in others it is due to hereditary endowments. When nature bases her selection upon acquired characters alone she fails to improve the average quality of the population. When she bases her selection upon the truly hereditary characters she effects a real, substantial improvement in the average quality of the population. In a single generation of selection the improvement may be only a slight one, but in the course of thousands of generations the average fitness of the surviving population may be markedly changed. This is evolution, for the final descendants may be so different from the original ancestors as to be regarded as a new species. And it is progressive evolution, for the descendants are better adapted to the environment than were the ancestors.

It should be clear, then, that evolution depends upon hereditary variation. It should be equally clear that natural selection does not produce this variation but merely guides it in the direction of better adaptation. Natural selection could accomplish nothing unless there were agencies at play to produce the hereditary variations with which she works. What are these agencies?

The original production of new hereditary variants is the phenomenon which biologists call "mutation." To give the phenomenon a name is not to explain it, and indeed biologists are even today far from understanding the real nature of mutation. They have found, however, that living organisms carry within the nuclei of their cells certain visible structures known as "chromosomes"—structures that are faithfully divided and passed on to the daughter-cells every time that cell division occurs—structures that are passed on via the gametes and the zygotes to the individuals of the next generation. They have found, by remarkable indirect methods, that these tiny but visible chromosomes carry many still smaller units which are themselves beyond the limits of visibility, even with the best of microscopes. These smaller units are the "genes" and it is they that are the fundamental units of heredity.*

Biologists have found that all characters of living organisms are due to the influence of these genes—interacting, of course, with the environmental influences that play upon the development of the organism. When one speaks of hereditary differences, he is referring to differences in the genes which two individuals possess. When one speaks of hereditary variations, he is referring to the fact that some of the genes differ among the individuals of a species. There are two great natural agencies that account for the gene differences between individuals. One of these is a primary agency, the other secondary. The primary agency is mutation.

Apparently the genes are well insulated against the influence of most environmental stimuli, for they are perpetuated without change for many generations. On rare occasions, however, some one gene, out of the thousands that are present in a nucleus, changes

---

* It is hoped that the student will pursue his biology far enough to make some study of that branch of the subject that is known as "genetics." As the result of a great body of experimental work, together with a remarkably clean-cut interpretation of results, genetics has provided a far more refined analysis of the machinery of heredity than is generally realized.

very suddenly to be a gene of a different sort. This change is the phenomenon of mutation.* The new gene will be perpetuated as faithfully as the old; so that mutation may lead to the presence in a population of several individuals that possess the new gene. The effect of the new gene will be to produce in those individuals either a brand new character (or characters) or a different grade or quality of development of some old character.

A given gene mutates only very rarely. There are apparently thousands of different genes within each nucleus, however; there are certainly millions of nuclei within the body of one of our higher organisms; and there may well be thousands or millions of individuals of the same species in the local population. It follows that the mutation of some gene somewhere in the population will be a fairly frequent event.

Apparently mutations are quite random in their occurrence. It appears to be a matter of chance which gene will mutate next or in what way the resulting new gene will affect the organism. Since mutations are random, we are not surprised to find that most of them have a bad effect upon the organism. For the organism is a complicated, intricate, finely balanced mechanism, and one would expect random changes to impair the efficiency of such a mechanism more often than to improve it.

If, now, most mutations have a bad effect, will not the phenomenon of mutation bring a steady deterioration of the race? Undoubtedly this would result were it not for natural selection. Actually, natural selection acts as a safety-valve, eliminating the bad mutations, preserving the good ones, and thus guiding the results of this purely random phenomenon in the direction of progressive evolution. If the new gene so modifies the organism as to make it more fit, the organism itself will probably survive, so that the gene in

---

* Actually, there are several known types of mutation. The commonest type, which is referred to above, is known as "gene mutation," and it is this type that has apparently played the largest rôle in evolution.

As for the cause of gene mutation, much remains to be learned. Penetrating radiation (such as X-rays) and perhaps high temperatures are stimuli that have been shown to induce mutation, but apparently nature includes other effective stimuli that have not as yet been identified.

question will be not only perpetuated but multiplied. If the new gene makes the organism less fit, both organism and gene will probably be exterminated through the action of natural selection, and the race will be no worse off than it was before.

In the course of many generations the new gene that improves the organism will be so effectively multiplied that all surviving members of the population will contain that gene. The resulting improved adaptation, slight though it may be, will be a substantial gain, and will provide a foundation upon which further improvements may be added in a similar way upon the introduction of additional new genes by mutation. Progress may be extremely slow, and there may even be long stretches of time through which little or no progress is made. In the long run, however, the accumulation of larger and smaller mutations will be sufficient to account for the origin of a new species.

Thus it is that modern biology explains evolution through random mutation plus natural selection. Adaptation in living organisms is the inevitable result of the operation of these forces of nature upon countless generations of ancestors.

Diversity in plant and animal kingdoms is due to the fact that different lines of descent from the same ancestor have often become separated from each other. Isolated in this way, the two lines may gradually diverge, not only because they may be living under slightly different environmental conditions (which may favor the survival of different genes), but also because the random mutations that occur in one line will probably differ from those that occur in another. Life-problems can often be solved in several different ways. Hence chance mutation might well be expected to launch one line of descent in the direction of one form of adaptation, and another (isolated) line in the direction of a different form of adaptation.

Earlier it was stated that there were two great natural agencies that accounted for the gene differences between organisms, i.e., for hereditary variations. It was further stated that mutation was the primary agency. The secondary agency is sex.

Other things being equal, the rate of evolutionary progress will depend upon the frequency with which new hereditary variations are introduced into a population. Evolutionary progress at a slow rate will result from mutation alone. Sex, however, has a way of increas-

ing still further the hereditary variation that is introduced by mutation. By thus increasing the frequency with which new hereditary
variations appear in the population, sex expedites evolution.

The student can develop a clear understanding of this action of
sex only through a knowledge of the machinery of heredity. It is
hoped that he will later gain this knowledge through a study of that
branch of biology that is known as "genetics." Here we can do no
more than to point out the general nature of the phenomenon.

Accumulation of mutations may bring into existence two varieties
of the same species which differ from each other by two genes. One
variety may possess genes a and B; while the other variety possesses
genes A and b. With asexual reproduction these two varieties will
perpetuate themselves faithfully, so that the species will continue to
include only these two varieties until such times as a new mutation
might occur. With sexual reproduction, however, the two varieties
may mate together. Under these conditions the machinery of heredity will operate to effect new combinations of the ancestral genes.
Among the descendants there will energe not merely the old combinations, Ab, and aB, but two new combinations ab and AB. Four
varieties now exist where there were but two before. Sex has not
produced variation in a homogeneous population, but it has multiplied a pre-existing variation that had been the result of mutation.

This simple example falls far short of showing how prodigiously
sex may multiply variation. If the two original varieties had differed
not by merely two genes, but by ten, sex could produce from these
two original varieties no less than 1,024 varieties. Would not that
combination of characters of the highest adaptive value be much
more likely to be found among the 1,024 varieties than among the
original two?

Nature has answered this question in the affirmative, for she has
preserved sex as the prevailing method of reproduction among our
highest plants and animals. As we have seen before, sex is not the
only method of reproduction; and of the several methods it is certainly not the most economical in perpetuating and multiplying the
species. It is the only method, however, which makes possible a
mixing of the hereditary characters from two parents, and is, therefore, the only method of reproduction that can lead to a multiplica-

tion of variation. So the student should not think of sex as *the* meth-od of reproduction, or merely as *a* method of reproduction; he should think of sex instead as *that* method of reproduction which has the peculiar power to multiply variation. This is the real function of sex, and it is a function which is of value only in making possible a more rapid evolution than might otherwise occur.*

---

* In textbooks of elementary botany it is conventional to describe in detail the distribution of chromosomes that occurs throughout the life-cycles of vari-ous plants. It is the opinion of the writer that such details are more happily presented in connection with a study of the machinery of heredity, in a context where they should be more meaningful to the student. Here we make no at-tempt to describe the machinery of heredity, for this small book is merely one of the several books that should properly make up the student's reading on introductory biology.

CHAPTER XX

CLASSIFICATION

FOR the organization of any branch of science, a primary essential is some serviceable system of classifying the materials which that science proposes to study. For biology this has been a tremendous task—a task which has not as yet been completed, down to its last detail, and never will be so completed. In remote quarters of the globe, many types of plants and animals are still waiting to be classified, and, even in our very backyards, evolution continues, slowly but surely, to bring new types into existence. No, there will never be a time when the "taxonomist" (classifier) can close his books with the statement that the last plant and the last animal has been identified, described, and assigned to its proper category. On the other hand, the past accomplishments of thousands of taxonomists, taken altogether, constitute an amazingly large contribution to human knowledge and provide a substantial foundation upon which the other branches of biology can build.

Most of us get experience with some sort of classification, from the boy who assembles his stamp collection, to the administrative officer who attempts to organize the machinery of his institution into some sort of workable hierarchy. In the main, this experience consists of arranging materials into categories of various magnitudes. Materials with much in common are put into the same small category. Small categories with somewhat less in common are assigned to a larger category. The larger categories themselves are assigned to a still larger category, and so on—until all the various materials or entities are grouped in a manner that reflects our opinion of their degrees of resemblance or difference.

If the materials to be classified are few, any one of several systems of classification may be adequate. If the materials are many, it is important to seek out the most understandable, workable, and serviceable system of classification and to adhere to that consistently throughout. In the early days of classification the several taxonomists used systems that were inconsistent, cumbersome, and not readily understood by their colleagues. The result was chaos. From this chaotic state modern taxonomy emerged through adopting (1) a consistent procedure and (2) a guiding concept.

For the consistent procedure credit is usually given to the great Swedish botanist, Linnaeus.* Between 1735 and 1758 Linnaeus published various classified lists of plants and animals. These were of some importance in themselves, being the best and most complete lists that had been published up to that time. Even more important, however, was their influence in establishing standards which were adopted by taxonomists all over the world. Casting aside the cumbersome descriptions that were employed by his predecessors, Linnaeus introduced a brief, graphic method of formal description for each type of plant and animal. (The description was written in Latin, which was agreed upon as the international language of science.) To each type he attached a formal Latin title which was to be regarded as the official scientific name of that type of organism all over the world and for all time to come. The title was a double one, and this system of "binomial nomenclature" has been employed ever since. The name of the genus is written first (and always capitalized), the name of the species second. This combination of generic and specific names makes it readily possible to assign to each known species a title that is absolutely distinctive. Thus, *Homo sapiens* can refer to only one thing, i.e., the species *sapiens* of the genus *Homo*, which happens to be the species to which all living men belong,† while *Trillium grandiflorum* refers only to a particular species of flowering plant that grows in the woods of North America.

So the process of settling upon the official title itself carried with

---

* Carl Linnaeus (1707–78) is commonly referred to as the "father" or "founder" of modern taxonomy. Actually, many of the ideas from which Linnaeus built his system were borrowed from various predecessors. His publications, however, were certainly the first to effect any general standardization of the science.

† The fossil record reveals other species of this genus, e.g., *Homo neanderthalensis.* But even where a genus is known to contain only one species among living and fossil forms, the full bionomial is employed. This is not only for consistency in practice, but provides for the possible subsequent discovery of other species of the same genus.

it the first step in classification, for it involved the assignment of similar species to the same genus. Beyond this, similar genera were assigned to the same "family," similar families to the same "order," and so on, culminating in the four great divisions of the plant kingdom (i.e., Thallophytes, Bryophytes, Pteridophytes, and Spermatophytes).*

Systematic arrangement of categories within categories is quite possible on the basis of any arbitrary criteria. One taxonomist, for example, might delimit great groups on the basis of flower size, smaller groups on the basis of flower color, and still smaller groups on the basis of number of floral parts; while another taxonomist might delimit great groups on the basis of number of floral parts, smaller groups on the basis of floral size, and still smaller groups on the basis of floral color. The final results of the two classifications would be conspicuously different. To a large extent this very state of affairs did exist until the biological world was provided with an underlying concept which helped to standardize the method of delimiting groups. The concept was that of organic evolution, effectively brought to the biological public by Darwin in 1859.

The guiding concept which evolution gives to classification is the concept of "blood relationship" or "common ancestry." An effort is

---

* It is sufficient at this stage for the student to visualize the essence of the system employed in classification. Memorization of the titles of groups of various rank can await the time that he decides to make practical use of the details of classification. Actually there are some inconsistencies in the application of the categories by various taxonomists. Species, genus, family, and order are always employed; but groups smaller than the species are variously referred to as "varieties," "sub-species," "races," etc.; while intermediate subdivisions of the larger categories are sometimes employed, such as "sub-genus" and "sub-family."

The concept of the magnitude of the different ranks also varies somewhat from taxonomist to taxonomist. Efforts have been made to recognize as constituting a species all organisms that can interbreed freely, but various practical difficulties have stood in the way of a rigorous application of this standard. Actually, concepts as to the appropriate breadth for species, genus, etc., have been essentially arbitrary, and there has often been difficulty in establishing a consensus of opinion on these matters. In spite of these difficulties, however, existing systems of classification are fairly well standardized and useful.

made to have classification represent degree of "blood relationship." All species thought to have descended from a relatively recent common ancestor are put into the same genus; all genera thought to have descended from a somewhat more remote common ancestor are put into the same family; and so on.

In making this effort, the following evolutionary considerations are kept in mind. It is clear that, in general, different lines of descent have diverged, not with respect to just one of their characters at a time, but simultaneously with respect to many. It is equally clear that the rate of divergence is less for certain "conservative" characters than for other less conservative, more "fluctuating" characters. The classifier judges relationship in terms of the totality of characters, but weighs more heavily those that his experience tells him are the more conservative. Thus, the presence of a seed is a very conservative character, and all plants with this character are thrown into the great division of Spermatophytes; while such characters as color of flower, size of flower, or size of entire plant are known to be so highly fluctuating that they are used merely to distinguish the several species of the same genus.*

The point has already been made that the four great divisions of the plant kingdom are not equally represented in the vegetation of modern times. In most situations, the picture is overwhelmingly dominated by members of the Spermatophyte division. Among Spermatophytes, two subdivisions are regarded as co-ordinate in the taxonomic sense, the Gymnosperms with their naked seeds, and the Angiosperms with their incased seeds. Of these two, the usually large-bodied Gymnosperms are represented by only about five hundred species today, while the much more diversified Angiosperms can boast a total of fully one hundred and thirty thousand species. Since it is this subdivision of Spermatophytes that now dominates the plant world, the student may be interested in a brief description of some of the more prominent Angiosperm families.†

---

* Or perhaps merely the several varieties of the same species, provided other differences are not also present.

† In the remaining pages an attempt is made merely to provide a glimpse of the vast assemblage of Angiosperms. Only a few of the more commonly en-

At the outset, one must recognize two taxonomically co-ordinate divisions of Angiosperms, the "Monocotyledons" and the "Dicotyledons." The titles refer to the primary and most reliable distinction between the two groups—a distinction based upon the nature of their embryos. The major, root-stem axis of the embryo carries lateral members known as "cotyledons" ("seed-leaves"). Among Monocotyledons, one cotyledon is present on the embryo, among Dicotyledons, two (Fig. 119). In addition to this primary distinction, there are other features which, with a few exceptions, will serve to distinguish the two groups. Monocotyledons have scattered vascular bundles (Fig. 84), "parallel-veined" leaves (Fig. 72), and floral parts usually in threes or multiples of three; while dicotyledons have their vascular bundles organized into a single great vascular cylinder (Fig. 82), have "net-veined" leaves (Fig. 72), and floral parts most commonly in fives or fours.

In numbers of species, the monocotyledon group is the smaller. Four of its families will be considered. Of all families in the entire plant kingdom, the one that is of greatest economic importance to man is the "grass family" (Gramineae). Here the flowers are exceedingly simple, having neither sepals nor petals, being surrounded by tough, leaf-like "bracts."* The flowers are arranged in a loose or compact cluster. The sporophyte body as a whole ranges from that of the tiny lawn grasses with horizontal underground stems to that of the towering bamboo, whose thin, erect stem may be almost one hundred feet in height. Economically, the bamboo is of tremendous importance in the tropics, where man has come to depend upon it for structures of all types. The small grasses of temperate regions pro-

---

countered families are mentioned at all, and those but briefly. Actually there are about three hundred recognized families of Angiosperms. It is not surprising that thousands of taxonomists have devoted their lives to identification and classification of the many and diverse types that are contained within this vast realm. Nor is it surprising that many an amateur makes the identification of plants his avocation. For the cultivation of such an avocation, the student is referred to one of the many existing "manuals" or "keys" for the identification of flowering plants.

* These bracts constitute the so-called "chaff" of wheat. Their obvious value to the plant is that of protection for the reproductive parts.

vide pasture and hay for our live stock, as well as lawns and golf links for ourselves. Together with the beet (a dicotyledon), it is the sugar cane that provides the world with its sugar supply. Greatest of all in economic importance are those grasses that we speak of as the "cereals."* From the time that agriculture was first started by prehistoric man its largest emphasis has fallen upon the raising of cereal crops, and this single activity has been extremely influential in molding human history and human culture. For those of us who live in temperate regions, wheat is the most important cereal, but corn, oats, rye, and barley should not be overlooked. In the tropics, rice is the principal food of hundreds of millions of human beings.

Widespread in the tropics, and of great utility to man is the "palm family" (Palmaceae.) Here an enormous cluster of simple flowers develops within a single bract. The sporophyte body is most commonly a tree, with unbranched trunk surmounted by a dense crown of large leaves. The coconut-palm, date palm, sago palm, and others, are put to innumerable uses by the natives of tropical regions.

Most typical and easily identified of all monocotyledons are the members of the "lily family" (Liliaceae.) Here the floral parts stand out conspicuously and symmetrically in threes or multiples of three, with the petals brightly colored and sometimes the sepals as well. The plant body is usually herbaceous, and is usually provided with a bulb or some other form of underground stem, which makes possible a rapid development of aerial parts at the opening of the season favorable to growth. Although this family gives us our asparagus and onion, it is better known for its beauty than for its usefulness. Most commonly encountered are the trilliums, the lily-of-the-valley, various true lilies, tulips, dog-tooth "violet," and the hyacinths.

Most advanced and specialized of all Monocotyledons is the "orchid family" (Orchidaceae). Greatest of all monocotyledon families in number of species, the orchids are actually rather rare plants. Such plants as the grasses dominate huge stretches of the earth's surface with a dense mass of vegetation that includes countless billions

---

* The cereals are those grasses that are cultivated for their "grains." Seedlike in appearance, the grain is technically a fruit, for it actually includes ovary wall as well as seed proper.

of individual bodies. Orchids, on the other hand, are scattered here and there as isolated individuals, quite difficult for the collector to find in temperate regions but more plentiful in the moist tropics. The orchid flower is characterized by an "inferior ovary,"* and by a strange specialization of one of the petals. This one petal, known as the "lip," becomes quite unlike the others, assuming an expanded, tubular or sac-like shape, and often being spurred. Actually, the orchid flower includes several peculiar details which have apparently evolved as adaptations to insect pollination. There is a remarkable specificity in the orchid-insect relations, such that one could almost say that for every orchid species there exists in the same locality a species of insect that is equipped with the particular size, shape, and instincts that are required for the pollination of that orchid. The sporophyte body is herbaceous, moderate in size, and often epiphytic. Much prized for their beauty, the orchids are actually of little economic importance, save for one Mexican form which is our source of vanilla.

Not included in the four families that we have just sketched are two Monocotyledons of considerable economic importance. The banana is a moderate-sized tropical tree with enormous leaves and a fruit that is a substantial part of the diet of millions of people. The pineapple, cultivated in the West Indies and Florida and in Hawaii, is a low plant with stiff, sword-shaped leaves that produces the remarkable, large, compound fruit of commerce.

Among Dicotyledons the taxonomist recognizes two great assemblages—the "Archichlamydeae," in which the petals are either quite separate from each other or are entirely lacking, and the "Metachlamydeae," in which the petals are fused through part of their length into a tube which surrounds stamens and carpels. This state of the

---

* When sepals and petals branch off from the main axis at a point below the base of the ovary, the botanist says that the flower is "hypogynous," being characterized by a "superior ovary." When sepals and petals branch off from the main axis at a point above the top of the ovary, the botanist says that the flower is "epigynous," being characterized by an "inferior ovary." This feature, along with others, is used in determining which families are "higher" and which are "lower." Among monocotyledons, only the orchids and two other families that we have not mentioned have evolved to the condition of epigyny.

petals is merely one of the several differences which exist between the two groups. A few of the outstanding families of Archichlamydeae will be mentioned first.

Several small families of the lower Archichlamydeae include our most common hardwood or deciduous trees. Simple, inconspicuous, wind-pollinated flowers are present, arranged in such characteristic "cone-like" clusters* as those that we notice during early spring on the cottonwood trees. Beside the cottonwood, other well-known forms are the elms, walnuts, hickories, oaks, chestnuts, willows, birches, beech, and poplars other than the cottonwood itself.

The "buttercup family" (Ranunculaceae) includes herbaceous plants characterized by five sepals, five petals,† numerous stamens, and numerous distinct pistils. Clematis, anemone, hepatica, marsh marigold, peony, spurred larkspur, and columbine are included here.

The "mustard family" (Cruciferae) includes herbaceous plants with a pungent taste, which is exaggerated in commercial mustard. The flowers have four sepals, four petals in one set, four long and two short stamens, and two carpels. Included here are the stock, sweet alyssum, candytuft, wallflower, watercress, horse-radish, mustard, cabbage, turnip, and radish.

One of the most beautiful, and at the same time useful, of plant groups is the "rose family" (Rosaceae), in which the flower suggests that of the buttercup. The plant body may be in the form of herb, shrub, or tree. In addition to the many roses themselves, the family contributes a large portion of our commercial fruits, including strawberries, raspberries, blackberries, peaches, apricots, plums, cherries, apples, pears, and quinces.

Largest of the Archichlamydeae families is that of the "legumes" (Leguminosae), characterized by its irregular flowers, highly adapted (like the orchids) to insect pollination, and by the ripening of the ovary into the well-known "pod." The legumes are of widespread distribution, with a plant body that ranges from a small herb to a tree of large size. Its better-known representatives are sweet pea, wisteria, lupine, sensitive plant, locust, honey locust, coffee-tree,

---

* This type of flower cluster is technically known as an "ament" or "catkin."

† Though in some members of this family petals are absent.

clover, alfalfa, pea, numerous beans, and the peanut; also the acacias, rosewoods, and many other tropical trees.

Most advanced of the Archichlamydeae, but not very conspicuous, are the Umbelliferae, characterized by an inferior ovary and an organization of flowers into flat-topped clusters.* The umbellifers include parsnips, carrots, and celery.

Well-known Archichlamydeae that are not included in the families mentioned above are the tulip-tree (yellow poplar), magnolia, linden (basswood), sycamore, maple, buckeye, box-elder, sweet-gum, tupelo (black-gum), violets, pinks, geraniums, nasturtiums, fuchsias, cotton, flax, hemp, currants, gooseberries, grapes, citrous fruits, tea, and the cacao-tree (which yields chocolate).

The "Metachlamydeae," with their tubular corollas, include the families of the highest rank, a few of which are mentioned below.

The "heath family" (Ericaceae) is characterized by two sets or cycles of five stamens each, so that the floral parts as a whole are arranged into five cycles. Woody shrubs for the most part, these plants grow most frequently in the cooler regions of the earth. Included are trailing arbutus, bearberry, heather, rhododendron, azalea, mountain laurel, wintergreen, huckleberries, blueberries, and cranberries.

The "mint family" (Labiatae) is characterized by a "two-lipped" corolla, square stems, opposite leaves, and an ovary conspicuously divided into four lobes. It includes basil, pennyroyal, lavender, mint, horehound, hyssop, savory, marjoram, thyme, balm, sage, rosemary, and catnip.

In the "nightshade family" (Solonaceae) one finds a conspicuous, regular, tubular corolla, with floral parts arranged in four cycles. Here we find the nightshade itself, red pepper, ground cherry, belladonna, jimson weed, potato (Irish potato), tomato, and tobacco.

Highest of all plant families is the Compositae, characterized by the organization of numerous small flowers into a head so compact as to resemble a single flower. Herbaceous plants, widely distributed and very numerous in temperate regions, the composites include dandelion, sunflower, goldenrod, thistle, beggar-ticks, blazing star,

---

* Known as "umbels."

daisy, aster, everlasting, rosinwood (compass plant), ragweed, cockle-burr, zinnia, dahlia, cosmos, marigold, chrysanthemum, sagebrush, burdock, and lettuce. It is interesting to note that this culminating family of the entire plant kingdom is of little commercial value to man and actually supplies a large portion of those "weeds" that interfere with his cultivation of other plants.

Well-known "Metachlamydeae" that are not included in the families mentioned above are: the coffee plant (which yields the coffee berry of commerce), cinchona (which yields quinine), sweet potato, olive, and the gourd fruits (watermelon, muskmelon, cucumber, pumpkin, squash).

INDEX

[Italic numbers represent illustrations; otherwise page numbers. Abbreviations: n. means footnote; ff. means the pages following.]